Carbon Nanotubes - Redefining the World of Electronics

Edited by Prasanta Kumar Ghosh, Kunal Datta and Arti Dinkarrao Rushi

Published in London, United Kingdom

IntechOpen

Supporting open minds since 2005

Carbon Nanotubes - Redefining the World of Electronics
http://dx.doi.org/10.5772/intechopen.87724
Edited by Prasanta Kumar Ghosh, Kunal Datta and Arti Dinkarrao Rushi

Contributors
Syed Awais Rouf, Abdul Mannan Majeed, Hafiz Tariq Masood, Mudassira Sarwar, Waseem Abbas, Zahid Usman, Jabulani I. Gumede, James Carson, Shanganyane P. Hlangothi, Md Ruhul Amin, P. Ramesh Kumar, Yeampon Nakaramontri, Apinya Krainoi, Jobish Johns, Ekwipoo Karnkornsurapranee, Victor J. Cruz-Delgado, José Manuel Mata-Padilla, Juan Guillermo Martínez-Colunga, Carlos A. Ávila-Orta, Janett Anaid Valdéz-Garza, Tajamal Hussain, Maria Tariq, Adnan Mujahid, Mirza Nadeem Ahmad, Muhammad Imran Din, Azeem Intisar, Muhammad Zahid, Gopal Sundar, Prasanta Kumar Ghosh, Kunal Datta, Arti D. Rushi, Ilias Belharouak

First published in London, United Kingdom, 2021 by IntechOpen
IntechOpen is the global imprint of INTECHOPEN LIMITED, registered in England and Wales, registration number: 11086078, 5 Princes Gate Court, London, SW7 2QJ, United Kingdom
Printed in Croatia

British Library Cataloguing-in-Publication Data
A catalogue record for this book is available from the British Library

Additional hard and PDF copies can be obtained from orders@intechopen.com

Carbon Nanotubes - Redefining the World of Electronics
Edited by Prasanta Kumar Ghosh, Kunal Datta and Arti Dinkarrao Rushi
p. cm.
Print ISBN 978-1-83881-184-6
Online ISBN 978-1-83881-185-3
eBook (PDF) ISBN 978-1-83881-186-0

For EU product safety concerns:
IN TECH d.o.o., Prolaz Marije Krucifikse Kozulić 3, 51000 Rijeka, Croatia, info@intechopen.com or visit our website at intechopen.com.

Meet the editors

Prasanta Kumar Ghosh is currently an assistant professor in the School of Physics, Dr. Vishwanath Karad MIT World Peace University (MIT-WPU), Pune (MS), India. He is an active researcher in the broad area of organic/inorganic functional materials, including organic field-effect transistors, inorganic/organic hybrids, and advanced sensors based on organic matter and SHI Irradiation. He has fifteen years of experience in the field of nanomaterials, microdevices, and instrumentation. He provides industrial and research solutions with low-cost automation and indigenous instrumentation. He has published fifty-two research papers in various international journals, books, and conferences.

Kunal Datta is currently an assistant professor in the Industrial Automation Division, Deen Dayal Upadhayay Kaushal Kendra (DDUKK), Dr. Babasaheb Ambedkar Marathwada University, Aurangabad (MS), India. He has fifteen years of research experience in instrumentation, automation, and electrochemical sensors (based on functional nanomaterials, namely single-walled nanotubes, porphyrins, organic conducting polymers, and nanoparticles). His research focuses on the real-time measurement of inorganic/organic air pollutants from the environment, the development of technical didactic tools, and industrial robotics. He has published fifty-four research papers in various international journals, books, and conferences.

Arti Dinkarrao Rushi is currently an assistant professor in the Department of Electronics and Telecommunication Engineering, Maharashtra Institute of Technology, Aurangabad (MS), India. Her areas of interest include nanotechnology, electrochemical sensors, functionalized materials, macrocyclic compounds, and the application of sensors for environment protection. She has published thirty-nine research papers in various international journals, books, and conferences.

Contents

Preface

Since the discovery of carbon nanotubes, they have gained all the attention of researchers over the world. In 1991, Sumio Iijima discovered carbon nanotubes (CNTs) while working on the synthesis of fullerene. Since then, CNTs have been explored in various fields due to their extraordinary electrical, mechanical, optical, and chemical properties. This book, *Carbon Nanotubes - Redefining the World of Electronics*, reviews and discusses CNTs, their electrical properties, and their utility in the electronics world.

This book is divided into eight chapters. The first chapter is an introductory chapter that explains the applicability of CNTs in the electronics field and gives the overall scope of the work. The second chapter discusses the incorporation of CNTs as building blocks into various electronic devices. The third chapter talks about the synthesis and purification of CNTs, as synthesis parameters have a huge impact on the electrical properties of CNTs. The fourth and fifth chapter discuss the possibility of CNT composites with natural rubber and elastomer. The sixth chapter examines the preparation and characterization of electrically conducting polymer nanocomposites with CNTs. The formed composite shows great promise for applications in electronics. The seventh and eighth chapter discuss CNT applications in energy storage devices and transistor preparation.

We are thankful to the authors and publishers for their support in the development of this book.

We hope that researchers and academicians worldwide who are interested in CNTs will find this book useful.

Prasanta Kumar Ghosh
School of Physics,
Dr. Vishwanath Karad MIT World Peace University (MIT WPU),
Pune, India

Kunal Datta
DDUKK,
Dr. Babasaheb Ambedkar Marathwada University,
Aurangabad (M.S.), India

Arti Dinkarrao Rushi
Department of Basic Sciences and Humanities,
Maharashtra Institute of Technology,
Aurangabad (M.S.), India

Section 1

Introduction to Carbon Nanotubes- Redefining the World of Electronics

Chapter 1

Introductory Chapter: Introduction to Carbon Nanotubes- Redefining the World of Electronics

Kunal Datta, Prasanta Kumar Ghosh and Arti Rushi

1. Introduction

There is no doubt that scaling down of Si-based complementary metal-oxide-semiconductors has reached its limit on performance with the count of MOS transistors in commercially available CPU crossing 15 billion [1]. Although, a 2020 report [2] states that according to INTEL, there is enough scope of higher transistor density with every possibility of existence of Moore's Law, a wide strata of material scientists believe in bringing about improvements through exploration of emerging materials [3].

Carbon and its allotropes have always garnered sincere attention of researchers across industries and academia. Due to presence of two electrons in the p-orbital of carbon, the bonding structure offers interesting versatility. So far, the electronic application with carbon-based materials is concerned, the sp^2 hybridized carbon materials have shown extreme prospect. Both graphene and carbon nanotubes, representing sp^2 hybridized form of carbon, have remained at the core of persuasion as alternatives of Si for electronic device applications. However, as graphene does not have any inherent bandgap, semiconducting CNTs find a certain edge over graphene.

The first report of CNT based field -effect transistor (CNT-FET) at Delft University and IBM [4, 5] was based on classic back gate geometry similar to Si-based FET. Within a very short period of time, ultra-high mobility to the tune of 80,000 cm^2V^{-1} s^{-1} [6] was demonstrated in semiconducting single-walled-carbon-nanotubes (SWNTs). Such high carrier mobility, was further correlated with electric current capacity of CNTs to the order of 109 A cm^{-2} [7], thermal conductivity of CNTs (at room temperature) upto 3,500 Wm^{-1} K^{-1} [8] and excellent mechanical strength [9]. The entire scientific community was gaining every confidence on CNTs as future substitute of Si. However, it took more than a decade for first CNT based computer to see the daylight [10, 11]. In spite of magnificence of CNTs are electronic materials, there were several bottlenecks that refrained this material from mass scale implementation through standard fabrication facilities. Continuous research efforts, through the entire period since first report of CNT-FETs, have cautiously addressed each challenge. The saga of such efforts is comparable to any epic as right from fabrication, processing hurdles, to implementation of CNTs on substrates, none of the step could be easily achieved and/or optimized.

As a matter of fact, the level of purity that is the requirement benchmark for semiconductor fabrication facilities, is not yet achieved with mass scale yield of semiconducting CNTs, thought to be the key constituent of 'beyond CMOS' [12] technology. Basically, the chiral angle of CNTs decide if they are of metallic and semiconductor nature. While metallic CNTs offer low bias ballistic transport [13], it is the semiconducting CNTs, particularly SWNTs, that are appealing because of intrinsic switching behavior when applied as active layer in FET structures. However, in contrast to metallic nanotubes, their semiconducting counterparts suffer from lower conductivity. Chemical doping can definitely be employed to alleviate this shortcoming; however, it needs consideration that dopant introduction in CNTs distort the sp^2 structure and leads to higher scattering and subsequent lowering of carrier mobility [14]. Peng et al. [15] have applied intrinsic form of CNTs to obtain a high-performance CNT FET employing a doping -free process. The authors were successful in growing high purity parallel arrays of CNTs directly on insulators, a remarkable achievement towards maintaining perfect sp^2 lattice of CNT. Synthesis efforts has, thus, came across exceptional spectrum of efforts to control the tube diameter, and improving semiconducting to metallic SWNTs ratio.

Dendritic networks of CNT films have also been employed as active layer in FET structures through solution methods [16]. In such cases, however, FET performance is seriously challenged by bundling induced lowering of mobility [17]. Another aspect of CNTs based electronic devices that warrant critical consideration is the formation of Schottky Barrier (SB) between semiconductor and metal [18]. Performance of devices is seriously challenged under such case as presence of SB severely limits the injection of carriers from metal electrode to semiconductor or vice-versa, a critical problem towards development of electronic devices. Javey et al. [19] and Zhang et al. [20] had demonstrated ohmic contacts using semiconducting CNTs using Pd and Sc to the valence band and conductance band of CNT respectively. Back gated FETs developed in such manner had exhibited near to ballistic conduction and barrier independent injection of electrons into conduction band of CNT at temperature down to 4.3 K. Another important milestone in CNT based device fabrication was self- aligned gate structure, an important requisite for large IC fabrication, demonstrated by Zhang et al. [20]. Next, the term 'complementary' in CMOS needs consideration, and an ideal CMOS circuit requires symmetric behavior by 'n' and 'p' devices. However, such condition is not followed by Si and most of the conventional semiconductor materials [21]. The present strata of semiconductor materials offer very high electron mobility if compared to hole mobility. And in this particular aspect, semiconducting CNTs are strikingly, miles ahead. As the conduction and valence band structure in semiconducting CNTs exhibit perfect symmetry near Fermi level, same carrier mobility for both charge carriers could be obtained. Such behavior, supported with perfect ohmic contacts possible with semiconducting CNTs lead to highly symmetric CMOS structure with CNTs [22]. Finally, one has to note two advantageous facets over CMOS fabrication - (i) that CNTs can be implemented in fabrication process without need of doping (as generation of carriers in dependent on contacts), relieving several costly steps of fabrication, and (ii) CNTs have been successfully demonstrated for pass-transistor logic (PTL), where signal has been demonstrated to be applied to any of the three terminals of FET [23, 24], that invariably indicates that paradigm shift in semiconductor industry is not far away.

Still, as published in a recent report [25], semiconductor fabrication industry is still working on some of the basic bottlenecks that includes – (i) difference in electronic properties of individual nanotubes, (ii) placing of nanotubes in circuits (one has to remember that although single CNT devices have outperformed Si based devices of same size; not only handling a single CNT at mass scale is a critical challenge, such devices are not capable of handling strong electrical signal) (iii) CNTs in bundle lower carrier mobility, (iv) requirement of purity in CNT yield (in the sense that semiconducting CNTs should have higher percentage of yield), (v) inability of exposing CNT deposited chips to plasma etching etc. At this point, Anthony Vicari, a material analyst at Lux Research, in Boston can be aptly quoted – "Historically, carbon nanotubes have shown impressive boosts in the lab that have been difficult to scale to a product" [26].

It is therefore quite pertinent that consideration of CNTs as future electronic materials, to be more precise, as substitute of Si and its counterparts, has enough scope to be encouraged while a significant spectrum of nooses need to be unfastened. Scientists from both industry and academia have a lot to reveal.

This volume looks forward to introduce readers to a cautious gamut of topics, right from synthesis and purification of CNTs to fault tolerance in CNT Transistors based multi valued logic. Readers will be able to generate ideas through elaborate discussions on CNT composites for electronic applications and illustration of CNTs as future energy storage device. The topics have been chosen so as to cater a material enthusiast at initial phase of research and an avid researcher in this field as well.

Author details

Kunal Datta[1*], Prasanta Kumar Ghosh[2] and Arti Rushi[3]

1 DDUKK, Dr. Babasaheb Ambedkar Marathwada University, Aurangabad (M.S.), India

2 School of Physics, Dr. Vishwanath Karad MIT World Peace University (MIT WPU), Pune, India

3 Department of Basic Sciences and Humanities, Maharashtra Institute of Technology, Aurangabad (M.S.), India

*Address all correspondence to: k2me@rediffmail.com

References

[1] Huawei Announces Mate 40 Series: Powered by 15.3bn Transistors 5nm Kirin 9000 [Internet]. Available from: https://www.anandtech.com/show/16156/huawei-announces-mate-40-series [Accessed: 22 October 2020]

[2] Moore's Law Lives: Intel Says Chips Will Pack 50 Times More Transistors [Internet]. Available from: https://singularityhub.com/2020/08/23/moores-law-lives-intel-says-chips-will-pack-50-times-more-transistors/ [Accessed: 23 August 2020]

[3] R. F. Service, Science 323 (2009) 1002.

[4] S.J. Tans, A.R.M. Verschueren, and C. Dekker, Nature 393, 49 (1998).

[5] R. Martel, T. Schmidt, H.R. Shea, et al., Appl. Phys. Lett. 73, 2447 (1998).

[6] T Durkop, SA Getty, E Cobas, MS Fuhrer, Extraordinary mobility in semiconducting carbon nanotubes. Nano Lett. 4, 35-39 (2004). doi:10.1021/nl034841q

[7] A Javey, P Qi, Q Wang, H Dai, Ten- to 50-nm-long quasi-ballistic carbon nanotube devices obtained without complex lithography. Proc. Natl. Acad. Sci. U. S. A. 101, 13408-13410 (2004). doi:10.1073/pnas.0404450101

[8] E Pop, D Mann, Q Wang, KE Goodson, HJ Dai, Thermal conductance of an individual single-wall carbon nanotube above room temperature. Nano Lett. 6, 96-100 (2006). doi:10.1021/nl052145f

[9] C Lee, X Wei, JW Kysar, J Hone, Measurement of the elastic properties and intrinsic strength of monolayer graphene. Science 321, 385-388 (2008). doi:10.1126/science.1157996

[10] F. Kreupl, Nature 501 (2013) 495.

[11] M.M. Shulaker, et al. Nature 501 (2013) 526.

[12] C. Macilwain, Nature 436 (2005) 22.

[13] Marcus Freitag, Carbon nanotube electronics and devices. In: Michael J. O'Connell, Editor. Book Carbon Nanotube: Properties and Applications. 1st Edition. CRC Press: 2006.p. 82-118. doi:10.1201/9781315222127

[14] R. Chao, et al. IEEE Trans. Nanotechnol. 4 (2005) 153.

[15] L.M. Peng, et al. AIP Adv. 2 (2012) 041403.

[16] Jana Zaumseil 2015 Semicond. Sci. Technol. 30 074001.

[17] C Biswas, YH Lee, Graphene versus carbon nanotubes in electronic devices. Adv. Funct. Mater. 21, 3806-3826 (2011). doi:10.1002/adfm.201101241

[18] E.H. Rhoderick, R.H. Williams, Metal–Semiconductor-Contacts, Clarendon Press, Oxford, 1988. ISBN:019859335X, 9780198593355

[19] A. Javey, et al. Nano Lett. 4 (2004) 1319.

[20] Z.Y. Zhang, et al. Nano Lett. 7 (2007) 3603.

[21] I Vurgaftman, et al. Appl. Phys. Lett. 89 (2001) 5815

[22] Z.Y. Zhang, et al. ACS Nano 3 (2009) 3781

[23] N.H.E. Weste, D. Harris, CMOS VLSI Design: A Circuits and Systems Perspective, 4th Edition, Addison Wesley, 2009. ISBN 10: 0-321-54774-8

[24] R. Zimmermann, W. Fichtner, IEEE J. Solid-State Circuits 32 (1997) 1079

[25] Katherine Bourzac. Carbon nanotube computers face a make-or-break moment. C&EN Global Enterprise 2019 97 (8), 22-25. DOI: 10.1021/cen-09708-feature1

[26] Chemical & Engineering News, ISSN 0009-2347

Section 2

Preparation, Purification and Characterization of CNT and CNT /CP Nanocomposite

Chapter 2

Preparation and Characterization of Electrically Conductive Polymer Nanocomposites with Different Carbon Nanoparticles

Víctor J. Cruz-Delgado, Janett A. Valdez-Garza, José M. Mata-Padilla, Juan G. Martínez-Colunga and Carlos A. Ávila-Orta

Abstract

Carbon nanoparticles possess a combination of high electrical and thermal transport properties, as well as low density and different morphologies that make them a good choice to reinforce plastics. Polymer nanocomposites offer great expectations for new and unexpected applications due to the possibility of changing their electrical/thermal behavior by adding nanoparticles while retaining the flexibility and processability of plastics. The possibility of electrical and thermal conduction in a polymer matrix with low amounts of nanoparticles brings opportunity for high demanding applications such as electrical conductors, heat exchangers, sensors, and actuators. Polyolefin nanocomposites offer a significant challenge due to their insulative nature and low affinity for carbon nanoparticles; due to the latter, new production tendencies are proposed and investigated.

Keywords: carbon nanoparticles, polymer nanocomposites, electrically conductive, ultrasound-assisted melt extrusion, thermal properties

1. Introduction

1.1 Carbon nanoparticles

From the discovery of cylindrical nanometric structures composed of one or several layers of carbon atoms similar to graphite by Iijima in 1991 [1], the scientific community embarked on a fascinating multidisciplinary career in the study, synthesis, characterization, and possible applications of these new carbon nanostructures, excited by the unusual combination of properties that these nanomaterials possess, among which the conduction of electricity and heat, low density, high mechanical resistance and morphology stand out. These nanoparticles have diameters in the range of 1 to 100 nm, lengths of 10 to 1000 nm. They can contain one, two or up to 100 layers rolled on each other with an equidistant separation of 0.34 Å [2–4]. Later, Novoselov and Geim [5] made an enormous contribution to science with graphene discovery, whose laminar crystalline structure is composed entirely of carbon atoms with an sp^2

hybridization, with a thickness of only one atom of carbon. Graphene has unusual properties between a metal and a superconductor and high mechanical, elastic, and chemical resistance. Therefore, graphene has been studied and proposed for various applications in electronic, aerospace, automotive, medical, and food industries [6–13].

Due to the ease of modifying its structure by incorporating other chemical elements, hybridization with functional groups, and decoration with organic molecules, carbon nanoparticle applications have been expanded enormously, leading to countless applications. For example, the miniaturization of electrical circuits composed of one or more carbon nanotubes, chemical and electromechanical sensors based on carbon nanotubes, the storage of hydrogen for fuel cells, the increase in charge capacity in batteries based on graphene or graphene nanoplatelets as well as the filtration capacity at the molecular level using graphene-based membranes, besides the reinforcement of polymeric matrices, to name only a few [4, 7, 11, 14, 15].

1.2 Polymeric nanocomposites

Materials science has been searching to generate new materials that possess a balance of properties, making them ideal for new and unexpected applications. Within this vast field are composite materials, which have a continuous phase (metallic, ceramic, or polymeric) and a discontinuous phase (filler or additive), which generally have high filler or additive contents of up to 70%, such as the case of titanium oxide (TiO_2) or carbon black concentrates in a polyethylene matrix, since both additives are used as pigments in the plastics industry [16, 17]. With the beginning of nanotechnology and the growing supply of different carbon nanoparticles, a new class of materials has emerged called polymeric nanocomposites whose advantage lies in using a smaller quantity of particles to modify the behavior of the host matrix or continuous phase.

Electroconductive polymeric nanocomposites were originally based on graphite derivatives, later carbon nanofibers, carbon nanotubes (mono or multilayer), and recently on graphene or graphene nanoplatelets, as well as a wide variety of combinations between these and other nanoparticles with different nature and morphology [8, 17–20]. In order to improve the electrical properties of these materials, combinations of carbon nanotubes have been made with graphite, graphene, clays, copper oxide, titanium oxide, silver nanowires, etc.; in all cases, the aim is to generate three-dimensional networks interconnected to facilitate the passage of electrons or phonons, to generate an electro/thermo-conductive material [21, 22].

In addition to providing the ability to conduct heat and electricity since they can exhibit the Peltier and Seebeck effect, [23, 24] such effects are beneficial in the development of thermoelectric materials, polymeric nanocomposites have also exhibited a noticeable improvement in mechanical properties, a barrier to gases, thermal stability [6, 9, 25, 26] as well as the ability to modify the electrical properties of the host matrix to generate materials for capacitors, electromagnetic and/or radiofrequency shields, have even allowed the development of metamaterials capable of modifying their refractive index, dielectric constant and/or Seebeck effect [27–29].

1.3 Polymeric nanocomposites preparation methods

There are different methods for preparing polymeric nanocomposites, where the main objective up to now has been to achieve adequate dispersion and distribution of carbon nanoparticles that allow modulating the properties of the resulting material. Because carbon nanoparticles are held tightly together by van der Walls forces, different ways have been sought to separate them individually to combine them with a polymer later and obtain a homogeneous polymeric nanocomposite. The main methods employed to achieve this are briefly described below.

1.3.1 Mixed in solution

In this method, the polymer is dissolved in a suitable solvent with the aid of magnetic, mechanical and/or heat stirring to facilitate complete dissolution of the polymer. The carbon nanoparticles are suspended in the same liquid (solvent) or a combination of them, and magnetic, mechanical, or ultrasonic stirring is applied to improve the dispersion of the nanoparticles. Subsequently, both solutions are mixed and kept under stirring, then the solvents are evaporated with heat or slowly in an extraction hood (the above will depend on the nature and reactivity of the solvent). Finally, the resulting material, usually a dark-colored powder, is compacted by applying pressure and heat to obtain a useful material. At the laboratory level, it is the most used method for research purposes; however, the large amount of solvents used makes its scaling at an industrial level unfeasible [30–32].

1.3.2 Polymerization in situ

In this method, one of the monomers or solvents used to obtain the polymer is mixed with the nanoparticles until a homogeneous dispersion is achieved; subsequently, the other reagents, including the corresponding catalysts, are added, and the polymerization reaction is carried out under the conditions of usual temperature and pressure. At the end of the reaction, the product obtained is purified, and the excess solvent is eliminated to recover the polymer formed with the incorporated nanoparticles. Given the complexity of this method, polyethylene's polymerization in the presence of carbon nanotubes at the laboratory level and of polyamide 6 with nanoclays at an industrial level has been successfully reported [20, 33, 34].

1.3.3 Melt mixing

This method is the most widely used at the laboratory level to obtain polymeric nanocomposites; it consists of passing the polymer and nanoparticles through a twin-screw extruder, whereby applying heat, the polymer melts and is transported by the screws that in turn impart shear forces to mix the components, in the different mixing zones that the extruder has. The mixture leaves the extruder, is cooled, and cut to obtain a polymeric nanocomposite. Due to its simplicity, this process can be easily scaled to an industrial level, in addition to the fact that it does not generate waste and does not use solvents [35].

1.3.4 Ultrasound-assisted melt mixing

Given the low affinity of polyolefins and in general of polymers for carbon nanoparticles, modifications have been made to the conventional melt mixing method by applying ultrasound waves in some specific sections of the extruder. It has been reported that this method can significantly improve the dispersion of nanoparticles of different nature and geometry, even with high nanoparticle content [36]. Different variants have evolved; the main difference being the mode of generation and application of ultrasound waves; conventionally fixed frequency ultrasound waves are generated, which are applied constantly or intermittently [37]. In another embodiment, the ultrasound waves are applied constantly, gaining a dynamic frequency sweep in a given interval [35, 38, 39].

There are other methods used for the production of polymeric nanocomposites, mainly at the laboratory level. Nevertheless, the choice of method will broadly define the level of dispersion and distribution of the nanoparticles within the polymeric matrix, and therefore the properties of the resulting material.

2. Methodology

In **Table 1**, the most outstanding reports in electro/thermo-conductive polymer nanocomposites of the last five years are presented to have a broader outlook on the subject. By their nature, polyolefins are good electrical insulators exhibiting

System	σ (S/cm)	Weight (%)	κ (W/mK)	Method of preparation	Ref
PS/SSWCNT[a]	$1.25x10^{5}$	75	0.30	Ball milling	[40]
PVC/CNT	$2.3x10^{-1}$	61	0.06	Drop casting	[41]
PP/MWCNT[b,c] PP/MWCNT[d] PP/MWCNT[e] PP/MWCNT[f]	$1x10^{-10}$ $1x10^{-7}$ $1x10^{-4}$ $1x10^{-3}$	8	—	Melt mixing	[38]
PS-LDPE/MWCNT	$2.9x10^{-3}$	1.5	—	Melt mixing	[42]
PVC/CNT	$2.4x10^{-2}$	25	—	Solution	[24]
HDPE/CNT[g] HDPE/CNT[h]	$2x10^{-4}$ $5.8x10^{-5}$	15	0.60 0.06	Melt mixing	[43]
PP/MWCNT[i,c] PP/MWCNT[d] PP/MWCNT[e] PP/MWCNT[f]	$1x10^{-5}$ $1x10^{-4}$ $1x10^{-3}$ $1x10^{-2}$	8	—	Melt mixing	[39]
PVC/SG-CNT[j]	$3.35x10^{2}$	66	0.18	Drop casting	[24]
LDPE/MWCNT	$2.38x10^{-2}$	5	—	Solution	[44]
LDPE/MWCNT LDPE/GNP	$2x10^{-2}$ $1x10^{-6}$	20	0.67 0.58	Melt mixing	[45]
PP/CNT[k] PP/CNT[l] PP/CNT[m] PP/CNT[n] PP/CNT[o]	$1.6x10^{-2}$ $9.56x10^{-1}$ $1.21x10^{-1}$ $1x10^{-3}$ $1.05x10^{-1}$	2	—	Melt mixing	[46]
mLLDPE/MWCNT	$2.8x10^{-4}$	10	—	Melt mixing	[47]
LDPE/G LDPE/SWCNT	$1.0x10^{-5}$ $8.3x10^{-5}$	3	—	Melt mixing	[48]
PP/SWCNT PP/B-SWCNT[p]	$1.21x10^{-1}$ $3.58x10^{-1}$	2	0.28	Melt mixing	[49]
PP/N-MWCNT[n]	$4 x10^{-2}$	5	0.28	Melt mixing	[28]

[a]*SSWCNT small-bundle-diameter-single-walled CNTs.*
[b]*PP MFI = 34 g/10 min.*
[c]*Melt extruded without ultrasound.*
[d]*Melt extruded with ultrasound fixed frequency.*
[e]*Melt extruded with ultrasound variable frequency.*
[f]*Melt extruded previously dispersed in gas phase.*
[g]*Solid.*
[h]*Foam.*
[i]*PP MFI = 1200 g/10 min.*
[j]*SG-CNT supergrowth-CNT.*
[k]*CNT, NC700.*
[l]*CNT, CNS-PEG.*
[m]*CNT, Tuball.*
[n]*CNT, N-MWCNT A1, Nitrogen doped.*
[o]*CNT, N-MWCNT IFW, Nitrogen doped.*
[p]*Boron doped SWCNT.*

Table 1.
Electric/thermal parameters of the most relevant polymer nanocomposites with carbon nanoparticles.

electrical conductivity in the order of 10^{-12} to 10^{-15} S/cm. As can be seen, different techniques have been used for the preparation of polymeric nanocomposites, achieving fascinating results. It can also be seen that the most popular preparation method is melt mixing, which, as mentioned above, is a versatile and easily scalable method. Another variant that can be observed is that depending on the polymeric matrix; the result will change; even more important is the concentration of nanoparticles used. Another aspect that should be highlighted is the modification or doping of the carbon nanoparticles, which slightly increases this property. Finally, as is known, polyolefins are thermal insulators, and their thermal conductivity ranges between 0.1 to 0.4 W/mK. Thermal conductivity has also shown sharp increases, as shown in Aghelinejad and Leung's reports and Paszkiewicz et al. [45, 50], where the matrix used was polyethylene.

3. Case of study

The motivation of present work was to perform a screening of several carbon nanoparticles to obtain polymeric nanocomposites with a better balance on properties such as electro/thermal conduction, mechanical and thermal stability. For this purpose, different carbon nanoparticles were selected. Their main differences lie in morphology (laminar versus fibrillar), structure (flat versus rolled layers), and functionalization (modified versus un-modified surface, i.e., CNT). Besides, the use of different polyolefins such as polyethylene and polypropylene, which bear significant differences in structure. On the one hand, polyethylene possesses a main chain almost free of pendant groups; meanwhile, polypropylene's main chain contains one methylene group each three carbon atoms. The best candidate is expected to be used to manufacture prototypes of thermistors (temperature sensors based on a change in electrical resistivity).

3.1 Materials and methods

In the following section, the preparation of polymeric nanocomposites in high-density polyethylene (PE) and polypropylene (PP) and their combination with four types of carbon nanoparticles (CNP) are presented and discussed. In all cases, a content of 20% wt/wt of each nanoparticle was used. The characterization results by thermogravimetric analysis, mechanical properties in tension and bending, electrical resistivity, and dielectric constant as a function of frequency and thermal conductivity are also presented. The resins used to obtain the polymeric nanocomposites were the following: high-density polyethylene (PE) Alathon H4620 with MFI of 20 g/10 min and density of 0.940 g/cm^3 provided by LyondellBasell (TX, USA), also polypropylene (PP) Formolene 4111 T with MFI of 35 g/10 min and density of 0.9 g/cm^3 provided by Formosa Plastics, (Tamaulipas, Mexico). The carbon nanoparticles used and their main characteristics are listed in **Table 2**.

The materials' processing was carried out in a Thermo Scientific model PRISM 24MC twin-screw extruder; the diameter of the screws is 24 mm with a length/diameter ratio of 40:1. According to the formulation, a controlled feeder for powders and another for the resin were used, which were previously calibrated to dose the required amount. The addition of the nanoparticles and the resin was carried out simultaneously in the extruder. A screw rotational speed of 100 rpm was used, a flat temperature profile of 180 and 200°C for the nanocomposites with PE and PP, respectively. Under these conditions, a production speed of 3.2–3.5 Kg/h was obtained. To improve the nanoparticle's agglomerates' dispersion and distribution, a device specially designed

Material	Density (g/cm^3)	SSA* (m^2/g)	Average length (μ)	Average diameter (nm)	Purity (%)	Supplier
CNT[1]	2.1	200	20	20	90	CheapTubes, Inc
MCNT[2]	2.1	110	20	20	90	CheapTubes, Inc
GNP[3]	2.1	600	2	—	97	CheapTubes, Inc
CB[4]	2.1	240	—	15	95	Cabot Corp.

SSA, Specific surface area.
[1]*CNT industrial grade.*
[2]*MCNT, Industrial grade modified CNT with -COOH contain 1.2% of COOH groups.*
[3]*GNP, industrial grade graphene nanoplatelets.*
[4]*Carbon Black, Vulcan XC72 grade.*

Table 2.
Characteristics and properties of the different carbon nanoparticles.

to irradiate the extruded material with ultrasound waves was coupled at the extruder exit. The device consists of a chamber with controlled temperature; inside, there is a 12.5 mm diameter titanium catenoid sonotrode (Branson Corp.) connected to a homemade ultrasound wave generator, which can generate ultrasonic waves in the range of 10 to 50 kHz, with a 750 W power [35, 38]. Finally, the material was passed through a water bath and cutter. Subsequently, each material was compression-molded to obtain a 15 X 15 X 0.2 cm plate, and a PHI press was used, a pressure of 20 Tn, with temperatures of 180 and 200°C for the nanocomposites with PE and PP, respectively. Specimens were cut for the characterization of the polymeric nanocomposites.

The characterization of the polymeric nanocomposites was carried out using the following analytical techniques. The thermogravimetric analysis (TGA) was carried out using a thermogravimetric analyzer from TA Instruments model Q500, using a sample of approximately 8 mg, a temperature range of 25–600°C, with a heating rate of 10°C/min and an inert atmosphere with nitrogen gas with a flow of 50 ml/min. The mechanical properties were evaluated in a universal testing machine, Instron model 1000, for tension tests in accordance with the ASTM D638 standard, using V-type specimens and a stretched speed of 50 mm/min and a load cell of 10 kN. The flexion tests were carried out according to the ASTM D790 standard using 12 X 1.25 X 0.2 cm specimens in 3-point bending mode; in both cases, five measurements were made, and the average value was reported. The electrical properties of resistance and capacitance were measured with an LCR analyzer in samples of 1 X 1 X 0.2 cm, both faces of the specimen were covered with silver paint, and a copper wire was placed as an electrode. The measurement was carried out at room temperature using a frequency range from 20 Hz to 2 kHz in increments of one decade; 5 measurements were made, and the average value was reported. The thermal diffusivity determination was carried out in a TA Instruments thermal diffusivity analyzer Discovery Xenon Laser Flash model (DXF-200). The analyzed specimen had circular geometry with 12.5 x 2 mm dimensions; both faces were coated with carbon paint and one of them with silver paint to ensure good contact with the temperature sensors; the measurement was carried out in triplicate at 25°C.

3.2 Thermal stability

The study of the thermal stability in electrically conductive materials is of great importance because when an electric current circulates through them, they can undergo heating and alter their behavior or ability to conduct electricity. On the other hand, this analysis makes it possible to determine the thermal stability of the materials and the amount of mass that they can lose due to the effect of temperature in a controlled atmosphere. It should be mentioned that if the atmosphere is air, thermo-oxidative degradation will occur. In **Figure 1**, the corresponding thermograms to the nanocomposites based on PE and PP are presented. While in **Table 3**, the specific data for the mass loss of $T_{5\%}$ and $T_{50\%}$ are shown.

It can be observed that PE exhibits a loss of mass from 330°C, while polymeric nanocomposites exhibit this loss at a temperature around 411°C, regardless of the type of nanoparticle used. It is important to note that the nanocomposite containing CB exhibits the highest thermal stability. For PP, degradation begins at a temperature of 370°C, while for polymeric nanocomposites occurs around 420°C, regardless of the type of nanoparticle used. In this case, nanocomposites based on CNT and MCNT exhibit the highest thermal stability of all.

Various reports in the literature suggest that carbon nanoparticles provide greater thermal stability or heat resistance to polymers in general due to a mechanism based on the formation of a carbonaceous layer and a tortuous path similar to a labyrinth on the surface of the material that prevents the release of combustion gases [19, 26]. This analysis is of great importance for flame retardancy applications in aeronautics, automotive, and textile industries and to determine the safety temperature that the material can support before molten and inflamed by the passage of an electrical current.

3.3 Mechanical properties

The mechanical properties of polymeric nanocomposites are of great interest because, as mentioned above, the addition of carbon nanoparticles can improve

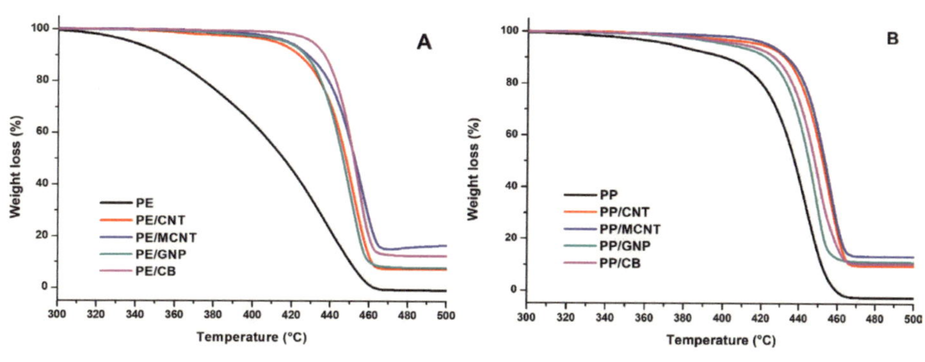

Figure 1.
Thermal stability by TGA of polymeric nanocomposites with 20% wt/wt of different CNP, (A) PE base, and (B) PP base.

Material	Polyethylene		Polypropylene	
	$T_{5\%}$	$T_{50\%}$	$T_{5\%}$	$T_{50\%}$
Polymer	337.50	415.67	373.49	437.59
CNT	411.28	447.81	420.09	451.99
MCNT	417.68	452.17	423.69	453.15
GNP	416.88	446.27	402.26	445.07
CB	430.27	451.89	419.47	449.98

Table 3.
Degradation temperatures at $T_{5\%}$, $T_{50\%}$, of polymeric nanocomposites with different carbon nanoparticles.

Material	Polyethylene			Polypropylene		
	Tensile modulus (MPa)	Elongation (%)	Flexural modulus (MPa)	Tensile modulus (MPa)	Elongation (%)	Flexural modulus (MPa)
Polymer	23.68	747	376	33.28	571	289
CNT	41.99	1	965	43.05	1	862
MCNT	38.84	1	989	38.7	1	800
GNP	42.03	1	1052	42.13	1	980
CB	40.47	1	951	44.22	1	913

Table 4.
Mechanical properties of polymeric nanocomposites with different carbon nanoparticles.

their performance. In **Table 4**, the properties of the PE and PP-based nanocomposites with the different carbon nanoparticles are listed.

As expected, with the addition of nanoparticles, the different properties were modified; firstly, the PE exhibits a tensile modulus of 23.68 MPa, while the nanocomposites present a maximum increase of 180%, this increase in resistance to stress causes the elongation of the material to be markedly reduced, suggesting that the stiffness of the material has changed from a ductile to a brittle material, in which plastic deformation has been suppressed. For its part, the flexural modulus corroborates the above since PE has a value of 376 MPa, and in nanocomposites, this value has increased to 280%. A similar behavior occurs with PP, exhibiting an increase of 130% and 330% in the tensile and flexural modulus, respectively. In this sense, the greatest increase in mechanical properties for polyethylene is obtained with GNP > CNT > CB > MCNT, while for polypropylene, it is CB > CNT > GNP > MCNT. In this sense, it is worth mentioning that the surface modification made to the MCNTs did not improve by itself, the compatibility with the host matrix PE or PP.

In the literature, many reports can be found that mention the improvement in mechanical properties in polymeric nanocomposites reinforced with carbon nanoparticles. However, the addition of compatibilizing agents such as maleic anhydride grafted to the resin is required to achieve a substantial increase in the mechanical properties, even with low amounts of carbon nanoparticles [9, 26, 51, 52]. Due to the lightweight and high modulus obtained by the polymeric nanocomposites reinforced with carbon nanoparticles, aeronautics and automotive industries would be benefited from the development of these materials for different components, which can provide a reduction in weight and lower consumption of fuels.

3.4 Electrical properties

The evaluation of electrical properties was carried out using an LCR as a function of a frequency interval, as shown in **Figure 2**. First, the polyethylene-based system allows observing that the PE resin exhibits the highest electrical resistance values at low-frequency values; above 10 kHz, the material becomes polarized and shows a lower electrical resistance, which decreases three orders of magnitude when reaching 2 MHz. With the addition of GNP, the material exhibits a behavior similar to that of PE, one order of magnitude lower in terms of electrical resistance. Meanwhile, the materials that contain MCNT and CNT show a reduction of 7 and 8 orders of magnitude; however, the polarization effect occurs when reaching high frequencies of 100 kHz. The CB-based system exhibits the least electrical resistance with nine orders of magnitude reduction concerning PE alone. In addition to not

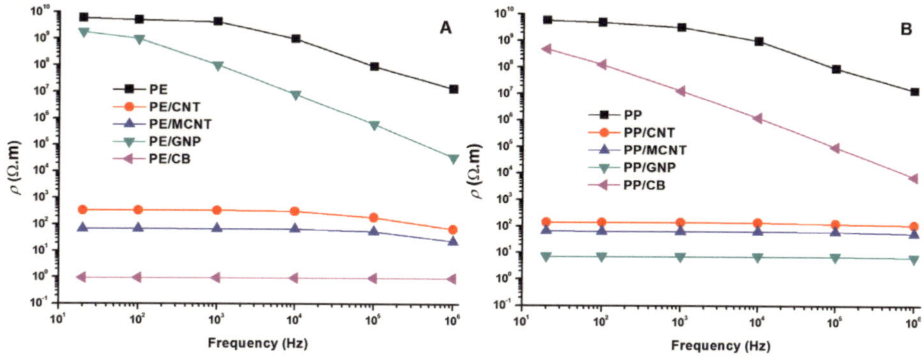

Figure 2.
Electrical resistance as a function of frequency, of polymeric nanocomposites with 20% wt/wt of different CNP, (A) PE base, and (B) PP base.

showing polarization effects as a function of frequency, which suggests that it behaves as an excellent electrical conductor.

For materials based on PP, the behavior is slightly different PP only presents the highest values of electrical resistance at low-frequency values; above 10 kHz, the material is polarized and shows a lower electrical resistance, which decreases three orders of magnitude when reaching 2 MHz, in the same way as the PE. Surprisingly, the CB-based system exhibits an electrical resistance that is completely dependent on the frequency. When it increases, the electrical resistance decreases to four orders of magnitude concerning the PP, suggesting that the material behaves like a semiconductor. On the other hand, the materials that contain CNT and MCNT show a reduction of seven and eight orders of magnitude without presenting the polarization effect in the entire frequency range, which suggests that they behave like a good electrical conductor. Finally, the compound containing GNP shows the lowest electrical resistance with a reduction of nine orders of magnitude and a linear response throughout the entire frequency range used. Based on the above, it can be pointed out that the nature of the polymeric matrix and the type of carbon nanoparticle can notably modify the electrical behavior of the polymeric nanocomposite [8, 31, 53, 54].

The behavior of the dielectric constant of polymeric nanocomposites is presented in **Figure 3**. Analogously to the behavior of electrical resistance, the dielectric constant follows a similar trend with the addition of carbon nanoparticles. The PE has a value of 3 and a linear behavior in the entire frequency range, while the nanocomposite with GNP shows an increase of 1 order of magnitude and a linear behavior as a function of frequency. Materials containing CNT and MCNT show an increase of three orders of magnitude for PE, with a slight decrease at high frequencies. The material that contains CB exhibits a frequency-dependent behavior since, at 20 Hz, it shows an increase of four orders of magnitude and then it decreases two orders of magnitude from a frequency of 1 kHz; this behavior corresponds to that of a capacitor, capable of storing energy and releasing it suddenly when used in electrical/electronic circuits.

On the other hand, PP exhibits a dielectric constant of 3 and does not vary as a function of frequency; the nanocomposite with CB shows an increase of one order of magnitude with respect to pure PP, while the nanocomposites with CNT and MCNT show an increase in 3 orders of magnitude and a slight decrease at high-frequency values. Finally, the nanocomposite with GNP presents the highest value of dielectric constant, with an increase of up to four orders of magnitude at a frequency of 20 Hz, and decreases by one order of magnitude for the rest of the frequencies evaluated. Similar to the behavior of PE nanocomposites,

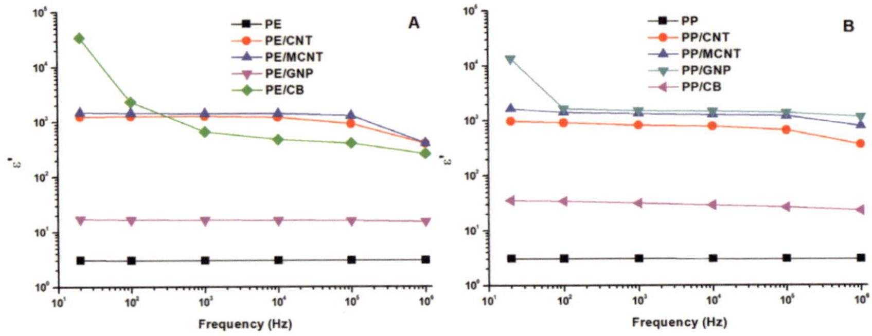

Figure 3.
Dielectric constant of polymeric nanocomposites with 20% wt/wt of different CNP, (A) PE base, and (B) PP base.

PP-based nanocomposites exhibit capacitor-like behavior throughout the evaluated frequency range.

The combination of properties for these new nanocomposite materials results in various applications that had not been previously conceived. For example, supercapacitors can be manufactured for systems that require a precise regulation of the supplied energy and a high energy storage capacity, and that in this way, the energy necessary to drive an electrical component can be supplied without the need to overload the electrical network of the circuit, besides not present a memory effect [25, 31]. Another field of interest for those materials would be the packaging industry, with the development of antistatic, static dissipative or semiconductive packages, for the protection of electronic components during their transportation, even for EMI or RF shielding for aerospace and defense to protect safety- and mission-critical systems from intentional and unintended electronics emissions [44]. The growing industry of electronic textile or smart textiles that develop wearable technology requires integrating textile fibers capable of conducting electrical signals. There are fabrics in which electrical and electronic elements such as microcontrollers, sensors, and actuators have been integrated that allow clothing to react, send information, or interact with the environment [55–57].

3.5 Thermal conductivity

The study of the thermal properties of polymeric nanocomposites intended for electronics applications is of great importance since, as mentioned above, the passage of electric current can induce a temperature gradient in electrical conductors, even in metals. The heat capacity was first determined, as well as the density and thermal diffusivity to determine the thermal conductivity of polymer nanocomposites. Values are shown in **Table 5**.

According to the data reported in **Table 5**, PE has the highest value of Cp; with the addition of the different nanoparticles, the Cp of the nanocomposites decreases significantly, the most notable case being the nanocomposite with CB. Meanwhile, PP exhibits an even higher Cp than PE, while the addition of the different nanoparticles promotes a decrease in this value, with graphene nanoplatelets being the material that most reduces this value. The decrease in Cp of the different nanocomposites can be associated with the ease they present for heat conduction, making the material less thermally insulating.

On the other hand, the thermal conductivity presents substantial improvements; in general, the PE-based nanocomposites exhibit the most significant increase in

	Polyethylene		Polypropylene	
	Cp (J/gK)	κ (W/mK)	Cp (J/gK)	κ (W/mK)
Polymer	1.846	0.24	1.917	0.28
CNT	1.671	0.43	1.672	0.32
MCNT	1.643	0.25	1.639	0.34
GNP	1.736	0.31	1.477	0.25
CB	1.495	0.28	1.569	0.30

Table 5.
Heat capacity (Cp, J/g K) and thermal conductivity (κ, W/m K) of polymeric nanocomposites with different carbon nanoparticles.

thermal conductivity 79, 29, 16, and 4% for the nanoparticles in the following order CNT > GNP > CB > MCNT, suggesting that carbon nanotubes are the most effective additive to increase the thermal conductivity of the nanocomposite. The trend is reversed, with increases of 21, 14, 7, and – 11% for MCNT > CNT > CB > GNP for PP-based nanocomposites. Although the Cp of the nanocomposites follows a different trend towards thermal conductivity, it should be mentioned that the type of polymeric matrix, the morphology, distribution, and dispersion of the different nanoparticles play an important role in heat conduction. This phenomenon is carried out through phonons; therefore, if there are spaces in the material in which the nanoparticles are too far apart, the phonons' passage through the material will find a physical barrier for their passage.

Recent reports suggest that a polymeric nanocomposite's thermal conductivity can be affected by different factors, including the processing method, the number of defects in the carbon nanoparticles, and, finally, their dispersion within the polymeric matrix [21, 29, 45, 46, 58]. The capability to conduct heat in a polymeric nanocomposite makes an ideal candidate for different applications such as heat exchangers, solar water heaters, thermoelectric materials, electrical heaters, to mention a few [22]. These devices will take advance of the lightweight, mechanical strength, thermal and dimensional stability of these materials, in which automotive, construction, and green industries are interested.

3.6 Thermistors

The electrical resistivity of polymeric nanocomposites with carbon nanoparticles shows an anomalous increase near the melting point of the matrix; this effect is known as a positive temperature coefficient (PTC) of resistivity. On the other hand, the negative temperature coefficient (NTC) is a very sharp decrease in resistivity when the temperature is above the melting point of semicrystalline polymers. These kinds of materials have important industrial applications like overcurrent protectors and self-regulating heaters [59, 60].

The polymer nanocomposites obtained were evaluated for their potential use as a thermistor. For this purpose, a prototype will be constructed; it consists of a square piece with dimensions 1 X 1 X 0.2 cm; both sides were cover with silver paste as an electrode and a copper wire. Kapton® tape was used to cover the prototype and isolate the wires during the heating cycle. A Mettler Toledo FP82 Hot Stage was used to supply heat in an interval from 40 to 160°C at a heating rate of 5°C/min, the Hot Stage was connected to a Mettler Toledo FP90 Central Processor, the electrical resistivity was measured with a Keithley Source Meter model 2400, in a 4-wire sense mode, to avoid the parasite signal in the circuit.

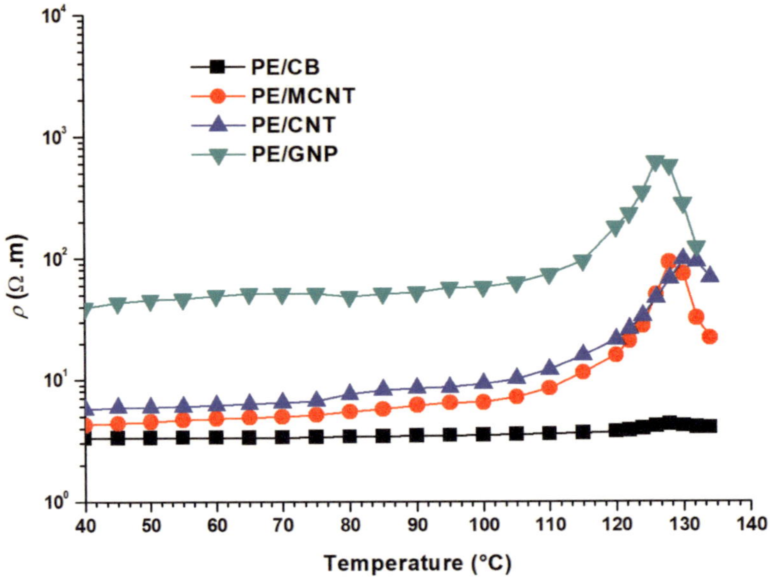

Figure 4.
Temperature versus electrical resistivity of PE base polymeric nanocomposites with 20% wt/wt of different CNP.

As seen in **Figure 4**, all the polymer nanocomposites exhibit thermistor behavior, i.e., an increase of resistivity around 128°C. The intensity of the PTC (the electrical resistivity ratio at the melting point versus room temperature) depends on the type of carbon nanoparticle used. The interval of temperature at which this phenomenon occurs is between 127 and 131°C. In this sense, the intensity of the PTC is in the following order GNP > CNT > MCNT > CB. This behavior could be associated with the capability of the polymer chains to break apart the conductive pathway formed in the polymer nanocomposite, due to the semicrystalline nature of the polymer matrix and the reduction in viscosity, during the heating. It is worth mentioning that PE/CB nanocomposite exhibits the lowest PTC intensity, probably due to the high structure of the CB (CB possess the small average particle size) and could form new conductive pathways in the molten state as stated by Zeng et al. [61].

4. Conclusions

The polymer nanocomposites with carbon nanoparticles become an electrically conductive material whit the addition of a certain amount of carbon nanoparticles; this property is fundamental in electrical and electronic applications. For many years, carbon black has been chosen as the best candidate for this purpose; with other carbon nanoparticles such as CNF, CNT, GO, graphene, and their combination with other materials, significant improvements have been made for electrically conductive materials.

In this work, the preparation and characterization of electrically conductive polymeric nanocomposites with different carbon nanoparticles was addressed to screen the type of carbon nanoparticles that allows them to obtain polymeric nanocomposites with a better balance on properties such as electro/thermal conduction, mechanical, and thermal stability. A material with the desired properties for their application in electronics, such as low electrical resistivity, thermal stability, and mechanical strength, besides thermal conductivity, is PE/CB polymeric nanocomposite since it exhibits a better balance of properties. This set of properties makes them candidates for use in various applications. Besides thermistors, they may be candidates for use in electrical heaters, which are a kind of electrical resistor used to converts electrical energy into thermal energy, as thermoelectric materials for their use in the exploitation of renewable energies, in heat exchangers, as EMI and RFI shielding, and as a wearable textile for smart applications.

Acknowledgements

The authors are grateful for the support of the CIQA technical staff for the preparation and characterization of materials: María G. Méndez Padilla, Gilberto F. Hurtado López, Rodrigo Cedillo García, Juan F. Zendejo Rodríguez and Jesús G. Rodríguez Velazquez. The financial support by SENER-CONACyT-CeMIE-SOL through the 207450-12 project is also appreciated.

Conflict of interest

The authors declare no 'conflict of interest'.

Author details

Víctor J. Cruz-Delgado[1*], Janett A. Valdez-Garza[1], José M. Mata-Padilla[2], Juan G. Martínez-Colunga[1] and Carlos A. Ávila-Orta[1*]

1 Centro de Investigación en Química Aplicada, Saltillo, Coahuila, Mexico

2 Consejo Nacional de Ciencia y Tecnología- Centro de Investigación en Química Aplicada, Saltillo, Coahuila, Mexico

*Address all correspondence to: victor.cruz@ciqa.edu.mx; carlos.avila@ciqa.edu.mx

References

[1] Iijima S. Helical microtubules of graphitic carbon. Nature 1991;354:56-8. DOI:10.1038/354056a0.

[2] Iijima S. Carbon nanotubes: Past, present, and future. Phys. B Condens. Matter, vol. 323, 2002, p. 1-5. DOI:/10.1016/S0921-4526(02)00869-4.

[3] Popov VN. Carbon nanotubes: Properties and application. Mater Sci Eng R Reports 2004;43:61-102. DOI:10.1016/j.mser.2003.10.001.

[4] Sgobba V, Guldi DM. Carbon nanotubes—electronic/electrochemical properties and application for nanoelectronics and photonics. Chem Soc Rev 2009;38:165-84. DOI:10.1039/b802652c.

[5] Novoselov KS, Jiang D, Schedin F, Booth TJ, Khotkevich V V., Morozov S V., et al. Two-dimensional atomic crystals. Proc Natl Acad Sci U S A 2005;102:10451-3. DOI:10.1073/pnas.0502848102.

[6] Cui Y, Kundalwal SI, Kumar S. Gas barrier performance of graphene/polymer nanocomposites. Carbon N Y 2016;98:313-33. DOI:10.1016/j.carbon.2015.11.018.

[7] Idumah CI, Hassan A. Emerging trends in graphene carbon based polymer nanocomposites and applications. Rev Chem Eng 2016;32:223-64. DOI:10.1515/revce-2015-0038.

[8] Khanam PN, Ponnamma D, AL-Madeed MA. Electrical properties of graphene polymer nanocomposites. Graphene-Based Polym. Nanocomposites Electron., 2015, p. 25-47. DOI:10.1007/978-3-319-13875-6_2.

[9] Papageorgiou DG, Kinloch IA, Young RJ. Mechanical properties of graphene and graphene-based nanocomposites. Prog Mater Sci 2017;90:75-127. DOI:10.1016/j.pmatsci.2017.07.004.

[10] Potts JR, Dreyer DR, Bielawski CW, Ruoff RS. Graphene-based polymer nanocomposites. Polymer (Guildf) 2011;52:5-25. DOI:10.1016/j.polymer.2010.11.042.

[11] Tan B, Thomas NL. A review of the water barrier properties of polymer/clay and polymer/graphene nanocomposites. J Memb Sci 2016;514:595-612. DOI:10.1016/j.memsci.2016.05.026.

[12] Zhang X, Samorì P. Graphene/Polymer Nanocomposites for Supercapacitors. ChemNanoMat 2017;3:362-72. DOI:10.1002/cnma.201700055.

[13] Zhang D, Zhang K, Wang Y, Wang Y, Yang Y. Thermoelectric effect induced electricity in stretchable graphene-polymer nanocomposites for ultrasensitive self-powered strain sensor system. Nano Energy 2019;56:25-32. DOI:10.1016/j.nanoen.2018.11.026.

[14] Zare Y, Shabani I. Polymer/metal nanocomposites for biomedical applications. Mater Sci Eng C 2016;60:195-203. DOI:10.1016/j.msec.2015.11.023.

[15] Verdejo R, Bernal MM, Romasanta LJ, Lopez-Manchado MA. Graphene filled polymer nanocomposites. J Mater Chem 2011;21:3301-10. DOI:10.1039/c0jm02708a.

[16] Tolinski M. Processing Aids for Extrusion. Addit. Polyolefins, 2009, p. 181-94. DOI:10.1016/b978-0-8155-2051-1.00012-1.

[17] Chaudhari S, Shaikh T and, Pandey P. A Review on Polymer Tio2 Nanocomposites. Int J Eng Res Appl

2013;3:1386-91. DOI:www.ijera.com/pages/v3no5#Paper232

[18] Kuila T, Tripathy T, Hee Lee J. Polyolefin-based polymer nanocomposites. Adv. Polym. Nanocomposites Types Appl., 2012, p. 181-215. DOI:10.1533/9780857096241.2.181.

[19] Sepahvand R, Adeli M, Astinchap B, Kabiri R. New nanocomposites containing metal nanoparticles, carbon nanotube and polymer. J Nanoparticle Res 2008;10:1309-18. DOI:10.1007/s11051-008-9411-2.

[20] Harito C, Bavykin D V., Yuliarto B, Dipojono HK, Walsh FC. Polymer nanocomposites having a high filler content: Synthesis, structures, properties, and applications. Nanoscale 2019;11:4653-82. DOI:10.1039/c9nr00117d.

[21] Gnanaseelan M, Chen Y, Luo J, Krause B, Pionteck J, Pötschke P, et al. Cellulose-carbon nanotube composite aerogels as novel thermoelectric materials. Compos Sci Technol 2018;163:133-40. DOI:10.1016/j.compscitech.2018.04.026.

[22] Dantas de Oliveira A, Augusto Gonçalves Beatrice C. Polymer Nanocomposites with Different Types of Nanofiller. In: Sivasankaran S, editor. Nanocomposites - Recent Evol., London, UK: IntechOpen; 2019, p. 103-28. DOI:10.5772/intechopen.81329.

[23] Dey A, Panja S, Sikder AK, Chattopadhyay S. One pot green synthesis of graphene-iron oxide nanocomposite (GINC): An efficient material for enhancement of thermoelectric performance. RSC Adv 2015;5:10358-64. DOI:10.1039/c4ra14655g.

[24] Oshima K, Inoue J, Sadakata S, Shiraishi Y, Toshima N. Hybrid-Type Organic Thermoelectric Materials Containing Nanoparticles as a Carrier Transport Promoter. J Electron Mater 2017;46:3207-14. DOI:10.1007/s11664-016-4888-4.

[25] Malas A, Bharati A, Verkinderen O, Goderis B, Moldenaers P, Cardinaels R. Effect of the GO reduction method on the dielectric properties, electrical conductivity and crystalline behavior of PEO/rGO nanocomposites. Polymers (Basel) 2017;9:613. DOI:10.3390/polym9110613.

[26] Venkata Ramana G, Padya B, Naresh Kumar R, Prabhakar KPV, Jain PK. Mechanical properties of multi-walled carbon nanotubes reinforced polymer nanocomposites. Indian J Eng Mater Sci 2010;17:331-7. DOI: http://hdl.handle.net/123456789/10508

[27] El-Shamy A gamal. Novel hybrid nanocomposite based on Poly(vinyl alcohol)/ carbon quantum dots/ fullerene (PVA/CQDs/C60) for thermoelectric power applications. Compos Part B Eng 2019;174:106993. DOI:10.1016/j.compositesb.2019.106993.

[28] Krause B, Konidakis I, Arjmand M, Sundararaj U, Fuge R, Liebscher M, et al. Nitrogen-Doped Carbon Nanotube / Polypropylene Composites with Negative Seebeck Coefficient. J Compos Sci 2020;4:14. DOI:10.3390/jcs4010014.

[29] Sun YC, Terakita D, Tseng AC, Naguib HE. Study on the thermoelectric properties of PVDF/MWCNT and PVDF/GNP composite foam. Smart Mater Struct 2015;24. DOI:10.1088/0964-1726/24/8/085034.

[30] Wiemann K, Kaminsky W, Gojny FH, Schulte K. Synthesis and properties of syndiotactic poly(propylene)/carbon nanofiber and nanotube composites prepared by in situ polymerization with metallocene/MAO catalysts. Macromol Chem

Phys 2005;206:1472-8. DOI:10.1002/macp.200500066.

[31] Xia X, Wang Y, Zhong Z, Weng GJ. A frequency-dependent theory of electrical conductivity and dielectric permittivity for graphene-polymer nanocomposites. Carbon N Y 2017;111:221-30. DOI:10.1016/j.carbon.2016.09.078.

[32] Moniruzzaman M, Winey KI. Polymer nanocomposites containing carbon nanotubes. Macromolecules 2006;39:5194-205. DOI:10.1021/ma060733p.

[33] Usuki A, Kojima Y, Kawasumi M, Okada A, Fukushima Y, Kurauchi T, et al. Synthesis of nylon 6-clay hybrid. J Mater Res 1993;8:1179-84. DOI:10.1557/JMR.1993.1179.

[34] Okada A, Usuki A. Twenty years of polymer-clay nanocomposites. Macromol Mater Eng 2006;291:1449-76. DOI:10.1002/mame.200600260.

[35] Cabello-Alvarado C, Reyes-Rodríguez P, Andrade-Guel M, Cadenas-Pliego G, Pérez-Alvarez M, Cruz-Delgado VJ, et al. Melt-mixed thermoplastic nanocomposite containing carbon nanotubes and titanium dioxide for flame retardancy applications. Polymers (Basel) 2019;11. DOI:10.3390/polym11071204.

[36] Ávila Orta CA, Martínez Colunga JG, Bueno Baqués D, Raudry López CE, Cruz Delgado VJ, González Morones P, et al. Proceso continuo asistido por ultrasonido de frecuencia y amplitud variable, para la preparación de nanocompuestos a base de polímeros y nanopartículas. Patent MX 323756B, 2014.

[37] Isayev AI, Kumar R, Lewis TM. Ultrasound assisted twin screw extrusion of polymer-nanocomposites containing carbon nanotubes. Polymer (Guildf) 2009;50:250-60. DOI:10.1016/j.polymer.2008.10.052.

[38] Ávila-Orta CA, Quiñones-Jurado Z V., Waldo-Mendoza MA, Rivera-Paz EA, Cruz-Delgado VJ, Mata-Padilla JM, et al. Ultrasound-assist extrusion methods for the fabrication of polymer nanocomposites based on polypropylene/multi-wall carbon nanotubes. Materials (Basel) 2015;8:7900-12. DOI:10.3390/ma8115431.

[39] Pérez-MedinaJC,Waldo-MendozaMA, Cruz-Delgado VJ, Quiñones-Jurado Z V., González-Morones P, Ziolo RF, et al. Metamaterial behavior of polymer nanocomposites based on polypropylene/multi-walled carbon nanotubes fabricated by means of ultrasound-assisted extrusion. Materials (Basel) 2016;9. DOI:10.3390/ma9110923.

[40] Suemori K, Watanabe Y, Hoshino S. Carbon nanotube bundles/polystyrene composites as high-performance flexible thermoelectric materials. Appl Phys Lett 2015;106. DOI:10.1063/1.4915622.

[41] Dey A, Hadavale S, Khan MAS, More P, Khanna PK, Sikder AK, et al. Polymer based graphene/titanium dioxide nanocomposite (GTNC): an emerging and efficient thermoelectric material. J Chem Soc Dalt Trans 2015;44:19248-55. DOI:10.1039/C5DT02877A.

[42] Patra R, Suin S, Mandal D, Khatua BB. Reduction of Percolation Threshold of Multiwall Carbon Nanotube (MWCNT) in Polystyrene (PS)/Low-Density Polyethylene (LDPE)/MWCNT Nanocomposites: An Eco-Friendly Approach. Polym Compos 2015;36:1574-83. DOI:10.1002/pc.23065.

[43] Aghelinejad M, Leung SN. Enhancement of thermoelectric conversion efficiency of polymer/carbon nanotube nanocomposites

through foaming-induced microstructuring. J Appl Polym Sci 2017;134:1-10. DOI:10.1002/app.45073.

[44] Gupta TK, Kumar S, Khan AZ, Varadarajan KM, Cantwell WJ. Self-sensing performance of MWCNT-low density polyethylene nanocomposites. Mater Res Express 2018;5:015703. DOI:10.1088/2053-1591/aa9f9e.

[45] Paszkiewicz S, Szymczyk A, Pawlikowska D, Subocz J, Zenker M, Masztak R. Electrically and thermally conductive low density polyethylene-based nanocomposites reinforced by MWCNT or hybrid MWCNT/graphene nanoplatelets with improved thermo-oxidative stability. Nanomaterials 2018;8:264. DOI:10.3390/nano8040264.

[46] Krause B, Barbier C, Levente J, Klaus M, Pötschke P. Screening of Different Carbon Nanotubes in Melt-Mixed Polymer Composites with Different Polymer Matrices for Their Thermoelectrical Properties. J Compos Sci 2019;3:106. DOI:10.3390/jcs3040106.

[47] Charitos I, Georgousis G, Kontou E. Preparation and thermomechanical characterization of metallocene linear low-density polyethylene/carbon nanotube nanocomposites. Polym Compos 2019;40:E1263-73. DOI:10.1002/pc.24961.

[48] Sabet M, Soleimani H, Mohammadian E. Effect of Graphene and Carbon Nanotube on Low-Density Polyethylene Nanocomposites. J Vinyl Addit Technol 2019;25:35-40. DOI:10.1002/vnl.21643.

[49] Krause B, Bezugly V, Khavrus V, Ye L, Cuniberti G, Pötschke P. Boron doping of SWCNTs as a way to enhance the thermoelectric properties of melt-mixed polypropylene/SWCNT composites. Energies 2020;13:394. DOI:10.3390/en13020394.

[50] Aghelinejad M, Leung SN. Fabrication of open-cell thermoelectric polymer nanocomposites by template-assisted multi-walled carbon nanotubes coating. Compos Part B Eng 2018;145:100-7. DOI:10.1016/j.compositesb.2018.03.030.

[51] Ferreira F V., Francisco W, Menezes BRC, Brito FS, Coutinho AS, Cividanes LS, et al. Correlation of surface treatment, dispersion and mechanical properties of HDPE/CNT nanocomposites. Appl Surf Sci 2016;389:921-9. DOI:10.1016/j.apsusc.2016.07.164.

[52] Wang B, Peng D, Lv R, Na B, Liu H, Yu Z. Generic melt compounding strategy using reactive graphene towards high performance polyethylene/graphene nanocomposites. Compos Sci Technol 2019;177:1-9. DOI:10.1016/j.compscitech.2019.04.013.

[53] Marsden AJ, Papageorgiou DG, Vallés C, Liscio A, Palermo V, Bissett MA, et al. Electrical percolation in graphene-polymer composites. 2D Mater 2018;5:032003. DOI:10.1088/2053-1583/aac055.

[54] Kuilla T, Bhadra S, Yao D, Kim NH, Bose S, Lee JH. Recent advances in graphene based polymer composites. Prog Polym Sci 2010;35:1350-75. DOI:10.1016/j.progpolymsci.2010.07.005.

[55] Nayak R, Wang L, Padhye R. Electronic textiles for military personnel. In: Dias T, editor. Electron. Text. Smart Fabr. Wearable Technol., vol. 1. 1st ed., Cambridge: Elsevier; 2015, p. 239-256. https://doi.org/10.1016/B978-0-08-100201-8.00012-6.

[56] Stegmaier T. Recent advances in textile manufacturing technology. In: Shishoo R, editor. Glob. Text. Cloth. Ind., vol. 1. 1st ed., Cambridge: Elsevier; 2012, p. 113-130. https://doi.org/10.1533/9780857095626.113.

[57] McKnight M, Agcayazi T, Ghosh T, Bozkurt A. Fiber-Based Sensors. In: Kai-Yu Tong R, editor. Wearable Technol. Med. Heal. Care, vol. 8. 1st ed., London: Elsevier; 2018, p. 153-171. https://doi.org/10.1016/B978-0-12-811810-8.00008-7.

[58] Prabhakar R, Hossain MS, Zheng W, Athikam PK, Zhang Y, Hsieh YY, et al. Tunneling-Limited Thermoelectric Transport in Carbon Nanotube Networks Embedded in Poly(dimethylsiloxane) Elastomer. ACS Appl Energy Mater 2019;2:2419-26. DOI:10.1021/acsaem.9b00227.

[59] Okutani C, Yokota T, Matsukawa R, Someya T. Suppressing the negative temperature coefficient effect of resistance in polymer composites with positive temperature coefficients of resistance by coating with parylene. J Mater Chem C 2020;8:7304-7308. https://doi.org/10.1039/D0TC00702A.

[60] Mamunya YP, Muzychenko YV, Lebedev EV, Boiteux G, Seytre G, Boullanger C, et al. PTC effect and structure of polymer composites based on polyethylene/polyoxymethylene blend filled with dispersed iron. Polym Eng Sci 2007;47:34-42. https://doi.org/10.1002/pen.20658.

[61] Zeng Y, Lu G, Wang H, Du J, Ying Z, Liu C. Positive temperature coefficient thermistors based on carbon nanotube/polymer composites. Sci Rep 2014;4:6684. https://doi.org/10.1038/srep06684.

Chapter 3

Synthesis and Purification of Carbon Nanotubes

Syed Awais Rouf, Zahid Usman, Hafiz Tariq Masood, Abdul Mannan Majeed, Mudassira Sarwar and Waseem Abbas

Abstract

In this chapter, we will evaluate the synthesis and purification of carbon nanotubes. Carbon nanotubes are cylindrical molecules that consists of graphene (rolled up of a single-layer carbon atom). A wide variety of synthesis techniques such as arc discharge synthesis, laser ablation of graphite/laser vaporization synthesis method, chemical vapor deposition (CVD), high pressure carbon monoxide synthesis and flame synthesis techniques, have been implemented to grow single and multi-walled carbon nanotubes for technological applications. All of the above methods exploit transition metals, like iron, cobalt, and nickel, as a catalyst. There are number of methods (filtering, chromatography and centrifugation) used to purify the carbon nanotubes, but the degree of purity remained questionable in these methods. In order to enhance the purification extent, alternate techniques such as Gas phase purification, Liquid phase purification and Purification by Intercalation are introduced. Here we will discuss the advantages and disadvantages of these purification routes. It will help researchers in selecting appropriate and effective method for synthesis and purification of carbon nanotubes.

Keywords: Graphene, Carbon Nanotubes, Synthesis, Purification, Laser Vaporization, Arc Discharge, Chemical Vapor Deposition, Gas Phase Purification, Liquid phase Purification, Intercalation

1. Introduction

Carbon atom contains six electrons with an electronic configuration of $1s^2$, $2s^2$ and $2p^2$. In its purest form, it crystallizes into graphite and diamond allotropes having different mechanical and optical properties. In former crystalline shape, the carbon atoms display sp^2 hybridization, where each carbon atom is covalently bonded with three other neighboring carbon atoms, making an angle of 120^0 in x-y plane along with a pi (π) bond available in z-direction [1]. This makes honeycomb like hexagonal crystal structure of graphene and this structural pattern shows the basis for other materials like fullerenes [2]. In diamond allotrope of carbon, carbon atoms unveil sp^3 – type of hybridizing character, forming a regular tetrahedron [1]. Apart from existing allotropes of carbon (diamond, graphite and fullerenes).

With the emergence of the field of nanotechnology, the carbon material (graphene, fullerenes and carbon nanotubes) where sp^2 hybridization prevails have attracted extreme focus from research community. Following the similar hierarchy, carbon nanotubes depict physical properties just like the graphene. Carbon nanotubes also offer sp^2 hybridization scheme and seems like a cylindrical coated graphene sheet in single and multiple wall patterns (**Figure 1a** and **b**). Nanotubes with single walls are named as single-wall carbon nanotubes (SWCNTs), firstly discovered in 1993 [5], while the nanotubes having multiple walls are termed as multi-wall carbon nanotubes (MWCNTs) discovered earlier in 1991 by Iijima et al. [6].

Immense interest in CNTs lies in their fascinating mechanical [7], electrical and optical properties [8] and hence are widely used in multiple applications such as field effect transistors [9], fuel cells [10], hydrogen energy storage applications [11], quantum computing [12] nanosensors [13–15] and battery electrodes [16]. The superior mechanical properties of CNTs are attributed to the higher values of tensile strengths and young modulus, thus revealing their potential use as a composite material to be used in futuristic Mars operation by NASA. Its use in such type of missions is subjected to its 50 times higher specific strength than the steel and hence creates exceptional load-bearing supports when integrated in composites. Field emission properties of CNTs have noticed enough attention from the research community, where the generation of electrons takes place under extreme conditions of electric field similar to thermionic emission. In addition, CNTs have also offered excellent chemical stability, higher electrical conductivity, nanosize and structural smoothness and are potentially used in flat display panels [8]. One can also attribute the use of CNTs in energy storage and energy production application to their smaller size, higher electron transfer rates, and superior surface topology in nanotubes.

As discussed above, CNTs have shown extremely smaller sizes, superior conductivity, greater mechanical strengths and elastic behavior, that is why these are used in other technological applications such as nanolithography, sensors, high resolution imaging and drug delivery systems also [17, 18].

Keeping in view the above-mentioned intriguing properties of CNTs, it is imperative to discuss the possible routes of their synthesis and the ways to

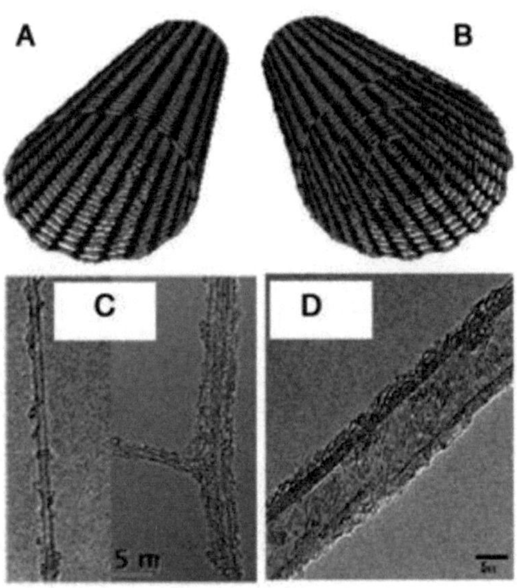

Figure 1.
Schematic representation of SWCNT (A) and MWCNT(B) along with the transmission electron microscope (TEM) images of (C) SWCNT and (D) MWCNT respectively [3, 4].

enhance purity of CNTs, as it will pave the way towards improved technological device applications.

2. Structural analysis of MWCNTs and SWNTs

The type of bonding among carbon atoms plays crucial role in determining its different allotropes with distinct physical properties. When carbon constitutes SP^2hybridization, a layered structure is formed with weak van der Waal forces existing in out of plane carbon atoms, in contrast to stronger in-planes bonding among them. Ideal CNTs can be thought of nano-scaled graphene cylindrical shapes closing at each end via half fullerenes. In case of multi-walled carbon nanotubes, there exist at least two equicentered cylinders of graphene and theoretically, these numbers of cylinders can be infinitely large. It should be noticed that there must be regular spacing between any two concentric grapheme cylinders in MWCNTs. Previous studies have demonstrated a real spacing width of the order of 0.34 to 0.39 nm [19].

The real space analysis of multiwall nanotube images has shown a range of interlayer spacing (0.34 to 0.39 nm).it has been observed that the inner diameter of such nanotubes varies from 0.4 nm to roughly few nanometers, in comparison to its outer diameter ranging from 2 nm to 30 nm. MWCNTs are closed from both ends by pentagonal type of ring defect named as half-fullerenes, with significant axial size difference (1 μm- few cm) between both ends [19].

Previous studies on SWNTs has documented their length 10^9 times greater than their diameters [20]. SWCNTs can be combined together in the form of ropes, to give hexagonal crystalline structure [21]. SWCNTs can assume three different types of structures such as armchair, chiral, and zigzag (**Figure 2B**) depending upon their wrapping in cylindrical form. The structure of SWCNTs is categorized by a pair of indices (n, m) that define chiral vector, which has prominent effect on the electrical properties of carbon nanotubes. Unit vectors along both directions in the crystal lattice is determined by the integers n and m. When

$m = 0$; nanotubes having zigzag structure.

$n = m$; nanotubes having armchair structure.

And other form is known as chiral structure.

The chiral vector can be defined as $C = na_1 + ma_2$ and it is used to measure the nanotube's diameter, where as a_1 and a_2 vectors explicate the grapheme base cell vectors. It is further stated that the chirality vector demonstrates the direction in which grapheme sheets are wrapped. One can estimate the diameter of a carbon nanotube is calculated by

$$d = a\sqrt{m^2 + mn + n^2} / \pi \qquad (1)$$

Where $a = 1.43 \times \sqrt{3}$ shows lattice parameters of the grapheme sheet.

When m = 0, we get zigzag CNTs and if m = n, one ends up with armchair CNTs. For other values of m, chiral CNTs will be formed. If the difference of n-m is a number which is multiple of 3, then the nanotubes will show metallic behavior and will be of highly conducting nature, otherwise one will be dealing with semiconducting or semimetal types of nanotubes. Armchair type of SWCNTs are metallic in nature, while other structures can make the SWCNTs semiconductor also. The Russian model and Parchment model are two broadly used models to prepare the

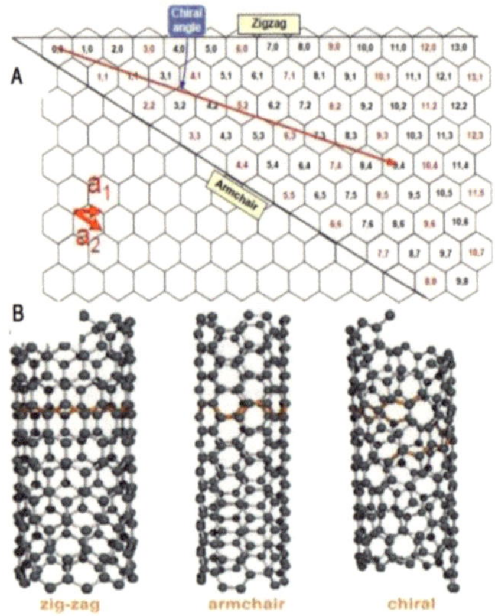

Figure 2.
Schematic of three different forms of SWNTs (A), where the chirality factor determines the diameter of carbon nanotubes and the (B) shows three different models of perfect SWCNT in atomic form [22].

MWCNTs. In the **Russian Doll model,** carbon nanotubes confine another nanotube inside and the diameter of the outer carbon nanotube is greater than the inner tube. When a single graphene sheet is rolled up many times just like the scroll of paper, it is known as **Parchment model.** The properties of SWCNTs are identical with MWCNTs. The outer layer in MWCNTs protects the inner CNTs against the chemical activity. It is considered as the main cause of higher tensile strength, which is absent in SWCNTs [23]. SWCNTs display sp^2 bonding between two independent carbon atoms, and hence result in higher tensile strength even compared to steel, when used as composite material. This sp^2 bonding is stronger than sp^3 bonding, present in diamond. CNTs show elastic behavior upon the application of a strong force. It bends and twists without undergoing permanent deformation in carbon nanotubes. When external force is removed, it will come back to its original form. Its elasticity is measured by a quantity known as modulus of elasticity (**Table 1**).

SWCNTS	MWCNTs
These are twistable but more flexible	These nanotubes cannot be easily twisted.
Its evaluation and characterization is relatively simple and easy.	It has very complex structure and hence their evaluation is not easy.
There are more chances of defect while working with SWCNTs.	The chances of defects are less but once occurred, are difficult to be removed.
Purity of SWCNTs is poor	Purity of MWCNTs is high.
Synthesis of SWCNTs on large scale is comparatively tough as it requires proper control over growth conditions.	Bulk synthesis is easy
Single layers are present in SWCNTs.	Multiple layers are available in MWCNTs.
The use of catalyst is compulsory for their synthesis.	MWCNTs can be prepared without using catalyst.

Table 1.
Difference between SWCNTs and MWCNTs [19].

3. Synthesis

There are multiple methods to synthesis CNTs where gas phase processes are involved. These methods are mainly known as arc-discharge synthesis technique, laser-ablation method and Chemical Vapor Deposition (CVD). Laser ablation method involves the synthesis of CNTs under high temperatures, while in arc discharge technique, the synthesis of CNTs occurs at relatively low temperatures (<800°C). CVD method is currently in use, as it allows the control of nanotube's length, diameter, alignment, density and purity with maximum accuracy [24] during the synthesis.

3.1 Synthesis of CNTs via arc discharge method

This method is implemented to synthesize the single and multi-walled carbon nanotubes (**Figure 3**) at a high temperature (above 1700°C).

The arc-discharge was initiated via applying a direct current of 200 A and a voltage of 20 V between the two electrodes. It was observed that the presence of iron, argon and methane was compulsory for the synthesis of SWNTs. The Arc discharge techniques is induced with the help of purest graphite electrodes having optical density of 6–10 mm and a diameter ranging from 6 to 12 mm. both of these electrodes were separated by 1–2 mm in a chamber containing helium gas at sub-atmospheric pressure. One can replace helium with hydrogen or methane gas. The working chamber consists of a graphitic anode and cathode, evaporated carbon [26] and minute amount of catalysts for example Ni, Co and Fe [27]. In arcing process is initiated by using direct current at pressure condition and the temperature of the chamber is raised up to 4000 K. In this procedure, half of the evaporated carbon is solidified on the tip of cathode. The rate at which evaporated carbon solidifies is set to be 1 mm/min and hence one gets "cigar like structure". During this process, the anode is also consumed. A remaining carbon is now a hard-gray shell structure, which is deposited on the edges and further condensates in the 'chamber soot' in nearby vicinity of the chamber's walls and 'cathode soot' on the negative graphite electrode (cathode). Furthermore, this inner material, anode soot and cathode soot (dark and soft materials) give rise to SWCNTS or MWCNTs along with nested graphene particles. Scanning electron microscopy (SEM) shows two different morphologies and surfaces are seen in the study of cathode deposited material. The

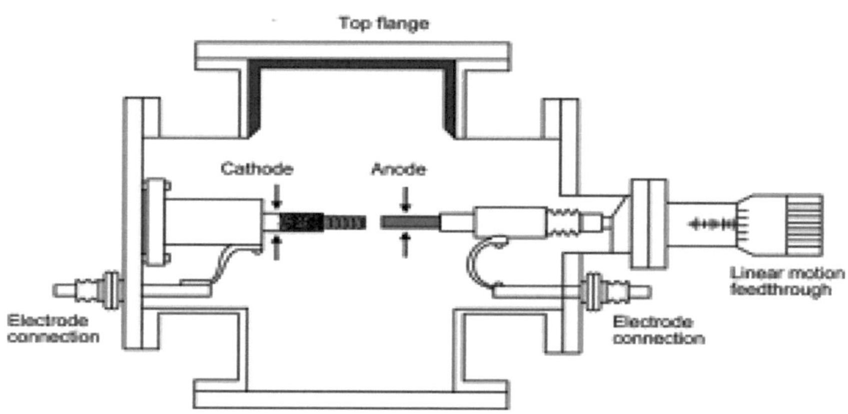

Figure 3.
The experimental set up of Arc discharge method [25].

soft and dark inner core contains randomly oriented carbon nanotubes and the grey colored outer core is composed of grapheme layers.

In arc discharge synthesis technique, there are two different options to synthesized the carbon nanotubes; one with and other without using the catalyst precursors. Generally, the synthesis of MWCNTs is performed without using catalyst precursors. On the other hand, the synthesis of SWCNTs is subjected to the presence of different catalyst precursors. In order to expand the arc discharge, a complex anode [28], made of metal and graphite, is exploited. The metals used in complex anode range from Fe, Ni, Pt, Pd, Co-Pt, Ag, to a mixture of Ni-Ti Ni-Y, Co-Ni, Co-Cu. It is demonstrated to get highest yield (< 90%) of SWCNTs by using a complex anode, made up of a mixture of Ni-Y with an average diameter size of 1.4 nm [29] and this mixture is utilized worldwide to prepare SWCNTs on a large scale. This method is considered one of the most practiced method to synthesize SWCNTS in large quantities. But the main disadvantage of this method is least control over the chirality in the intended nanotubes.

3.2 Laser ablation technique in the synthesis of CNTs

A graphite block is heated in quartz tube via high power lasers in a furnace at a temperature of 1200°C in argon atmosphere [30]. Here the laser vaporizes the graphite target within the quartz tube and SWCNTs are formed in the presence of metallic catalysts. The diameter of prepared carbon nanotubes is manipulated as a function of laser power such as the diameter of the tube decreases upon increasing the power of laser pulses and vice versa. Some other studies have dictated that the ultrafast sub-picosecond lasers have the ability to produce single walled carbon nanotubes on a large scale too [31]. It is further reported to manufacture carbon nanotubes up to 1.5 g/h via laser ablation method.

To harness CNTS with desired structural and chemical features, one should monitor the effect of different properties of lasers (peak power, frequency, oscillation wavelength, cw versus pulse), chamber pressure, distance between graphite target and substrate, ambient temperature and the flow and pressure of the buffer gas. By using this process, one can achieve high quality (purity) SWCNTs in large quantities. The mechanism and principles of laser ablation is identical with the arc-discharge method. Here the required energy is provided by a laser which strike with pure graphite pellet holding catalyst material i.e. cobalt and nickel (**Figure 4**).

The primary advantages of this method are the presence of the smaller amounts of metallic impurities and higher yield of CNTs. On the other hand, the carbon nanotubes produced via laser ablation method are not perfectly straight and uniform. This is an expense method due to the requirement of high purity graphite rod and the availability of two laser beams to produce CNTs. By using this technique, the yield of nanotubes per day is relatively smaller than the arc discharge technique.

3.3 Chemical vapor deposition for CNTs synthesis

One of the best techniques for the production of CNTs is chemical vapor deposition (CVD). There are different CVD techniques such as catalytic chemical vapor deposition either thermal [33] and water assisted [6], plasma enhanced oxygen assisted CVD [34–36] or hot filament CVD (HFCVD) [37]. But most extensively implemented CVD method for the production of CNTs is known as catalytic chemical vapor deposition. This route involves the Chemical breakdown of hydrocarbon on a specified substrate and helps expand the CNTs on different type of materials. Carbon atoms remain intact with the metallic catalytic particles, as was the case for arc discharge technique. After that carbon atoms are enabled to come in contact

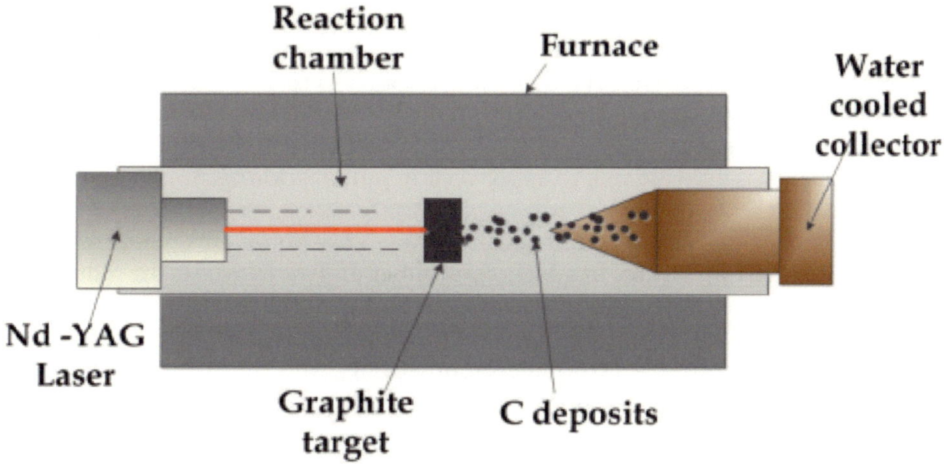

Figure 4.
The laser ablation process [32].

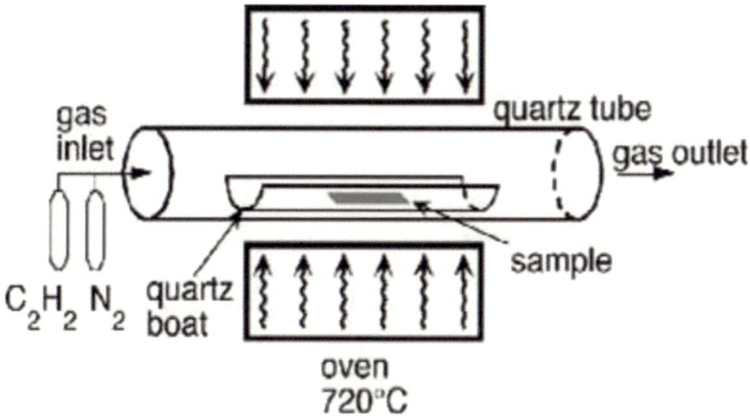

Figure 5.
Chemical vapor deposition [38].

with metal particles and implanted with in the holes, initiating the production of carbon nanotubes (**Figure 5**).

This technique facilitates well aligned long carbon nanotubes and a layer of metallic catalyst particles are produced at 700°C. Most commonly catalyst metals are cobalt, nickel, iron and combination. The expansion of nanotubes carried out in fluidized bed reactor in the presence of a gas containing carbon such as ethylene, acetylene, methane, etc. and a process gas like H, Ne, or ammonia are used as well. The process gas reacts with the catalyst particles and disintegrates. Carbon atoms become prominent at the edges of nanoparticles where CNTs are created. CVD is very economical practical method for quite pure and large-scale production of carbon nanotubes as compare to laser ablation method. This method is easily controllable and give high purity of obtained materials, this is the main advantages of CVD [39].

4. Purification of carbon nanotubes

Above mentioned as-synthesized methods of CNTs encounter certain impurities, such as smaller fullerenes, wrapped graphite sheets, metal catalyst particles, and amorphous carbon contaminations. It is observed that the percentage of these impurities generally enhances as long as the diameter of CNTs increases. Therefore, it is important to get rid of these impurities to obtain homogenously distributed CNTS in polymer or dispersion media due to their substantial effect on electro-mechanical properties of CNTs, interfering with the expected applications. It makes it unavoidable to apply certain purification techniques to get pure CNTs with better electrical and mechanical features [40, 41]. Due to the insoluble nature of CNTS, it is quite challenging to use liquid chromatography to get rid of these impurities. In addition, number of groups across the globe just characterize the commercially synthesized carbon nanotubes and do not have facilities to grow them. Due to the application of different analytical techniques such as Raman, scanning electron microscopy (SEM) and transmission electron microscopy (TEM), even SWCNTs have shown doubled, triple and multi-walls of a single sample along with the presence of above-mentioned impurities. Hence, one cannot rely on the specification provided by different companies. Subjected to these various analytical characterization and impurities, researchers have applied various purification techniques, leading to significant loss of CNTs [42–44]. It is further observed that the use of acid treatment or surfactants might result in CNTs with activated surfaces, putting comprehensive changes in their desired properties [45].

Depending upon the nature of the structure (single-walled or multi-walled) in hands, growth process, and metal catalysts, various purification techniques such as mechanical, chemical and physical routes are to get dispersed carbon nanotubes with maximum possible exclusion of impurities [23]. The chemical methods allow the variation in surface energy by introducing functionalization of carbon nanotubes. It leads to better wettability and adhesion of CNTs to the polymer target media and hence the tendency of agglomeration decreases. But the use of acids might deteriorate the structural quality of CNTs, attributing un-desired physical properties.

The chemical route of purification produces highly pure CNTs but fragile to structural defects and product losses [46]. However, CNTs with higher purity can be achieved by removing the metal catalyst particles in controlled reaction. Physical methods are attractive due to the possibility of adsorption of variety of functional groups, leaving behind similar pi (π)- graphene structure and are implemented when higher weight fraction of CNTs is desired. Physical method separates the yield products on the bases of the size of CNTs [47]. Physical methods. These methods cause low damages and are more complex as well as less effective as compared to chemical methods. Here we will only explain the chemical methods for purification [48]. The most commonly used chemical purification method involves oxidation of synthesized CNTs in gas phase as well as in liquid phase. Most common purification methods with high success rates are

- Gas Phase
- Liquid Phase
- Intercalation Method

4.1 Gas phase

Purification can be done in dry gas oxidation. Carbon dioxide, hydrogen gas and dry/wet air are commonly used oxidation gases for this method [49–51].

4.1.1 Air

Air oxidation is one of the gas phase oxidation methods to purify CNTs. The impurities in CNTs are removed by the thermal air oxidation at moderate temperatures. The walls of the CNTs and the binding between the entangled CNTs are affected by the presence of oxygen. It is also known as a strengthening process which starts at 480°C and amorphous carbon usually decayed between 480 and 500°C [52, 53]. The reactivity rate is greater for structure and amorphous carbon than cylindrical wall of CNTs when oxidation is done in air. Due to this selective oxidation, the amorphous carbon can be bare-off from the cross-linked CNT collections. If the temperature is raised to 750°C during the annealing process, the loss rate of CNTs enhances to about 90% and the structure of CNTs is destroyed significantly.

4.1.2 Other gases

As in gas phase oxidation methods, controlled rate heating of CNTs is implemented for a longer period of time. Here the disordered amorphous carbon that is coming from the tip, destructs the purification on the base of oxidation by CO_2 . Mild oxidation carried out with CO_2 at 600°C [54].

The route of the reaction is shown:

$$CO_2 + C \rightarrow 2CO \tag{2}$$

Amorphous carbon and metal catalyst particles coated with carbon may be removed by hydrogen gas treatment at high temperatures. Amorphous carbon is converted into the carbon dioxide in air and then transformed again into methane in the presence of hydrogen. Ammonia (NH_3) is used to remove residual carbon impurities and repair the damaged in sidewalls of CNTs, instead of using hydrogen gas. Ammonia has advantages over hydrogen in terms of ease in handling. Only a small number of defects are observed in the CNTs when they were exposed to NH_3 gas at high temperatures during purification process. In addition, strong van der wall forces are induced between CNTs after NH_3 treatment, leading to a damage recovery of sidewall [52, 55] of CNTs.

4.1.3 Effect of gas phase oxidation

Oxidation of amorphous carbon in gas phase is easy to control as compared to liquid phase oxidation techniques. Higher activation energies are required in gas phase oxidation processes. Gas phase oxidation can better oxidize the CNTs than the liquid phase oxidation without introducing defects. This yield purified nanotubes, arranged in tight bundles without forming clusters. Moreover, there is no need to use complicated/sophisticated equipment, filtration and separation processes required after the purification [56, 57].

4.2 Liquid phase

Despite the fact that the advantages of gas phase oxidation are clear, it has some limitations. Metal particles cannot be eliminated straightaway, and further treatment is compulsory. In order to overcome this drawback, liquid phase purification treatments are developed to eliminate the amorphous carbon and metal catalysts [57, 58].

The oxidant and mineral acid, in the form of solution can uniformly react with the network of the raw CNTs samples. Therefore, processing with selective oxidizing agent with precise control can yield high-purity CNTs. The scientific community mostly uses HNO_3, NaOH and H_2O_2 as an oxidizing agent for liquid phase oxidation.

4.2.1 Nitric acid oxidation

Nitric acid is commonly used for purification of CNTs due to its capability of removing metal catalysts, nontoxicity and economy. It can remove the amorphous carbon selectively because of its mild oxidizing ability. A concentrated nitric acid is used to produce SWCNTs through laser ablation in a single step. The synthesized SWCNTs were sonicated in concentrated nitric acid for a few minutes, following the refluxing for 4 h under magnetic stirring process carried at 120–130°C. The product reached 30–50 wt. % of its raw samples and the metal defects were reduced up to app. 1@ wt. %. The purity of SWCNTs and its production totally depends upon the concentration of nitric acid and reflux time in nitric acid treatment. The elimination of metallic impurities can be confirmed via XRD analysis of CNTs. During the purification, the nitric acid reacts with the defected parts and intercalate into the CNTs to unzip the tube walls by further oxidative etching, which in turn causes an increase in nanotubes inter-layer spacing. Normally, the reactive carbons were eliminated through the following chemical reaction:

$$C + 4HNO_3 \rightarrow CO_2 + 2H_2O + 4NO_2 \qquad (3)$$

Most of the catalyst particles are removed in nitric acid's treatment at high temperatures for 24 h. The unwanted impurities are removed and melted effectively from CNTs, and some oxidative defects in the sidewall of CNTs are also induced in this process. The intensity of the D-band spectrum produced by the defects and carbon particles in the sample after acid treatment, can be used to determine the disorder degree of the sample [57, 59].

4.2.2 Sodium hydroxide treatment

It has been observed clearly from Scanning electron microscope (SEM) that silica and alumina support can be eliminated significantly after NaOH treatment. Based on the reaction between NaOH and carbon, a single-step method for simultaneous purification and opening of multi-walled carbon nanotubes has been formulated [48, 60]. The redox reactions between carbon and NaOH followed through the highly reactive sites of the material. As a result, NaOH only interacts with the carbon impurities and defects of the carbon nanotubes, that is with the tip while the uniform graphite layers of the nanotubes walls remain intact. This is because metallic sodium cannot be inserted into well-organized materials, and can only be carried out by highly disordered carbon impurities [61]. Therefore, in addition to the opening of tubes, NaOH treatment removed the catalytic support, amorphous carbon, and the catalyst metal particles. Its mild conditions removed the metal impurities without damaging the sidewalls of CNTs.

4.2.3 Hydrogen peroxide oxidation

Hydrogen peroxide (H_2O_2) attacks on the carbon surface and cannot eliminate metal particles due to its mild oxidization capability. It is inexpensive and green oxidizing agent and is commonly used with HCl. Generally, H_2O_2 can be transformed into a toxic salt. H_2O_2 with HCl has been examined to eliminate the metallic particles during purification of CNTs. Macro-scale purification consists of two parts such as refluxing treatment in H_2O_2 solution following the cleaning process performed with HCl. Particle size of Fe has a significant effect on the oxidation of amorphous carbon. The oxidation and removal of metal particles in this process is performed in a single container to make it simple. The purity and yield of the product in this treatment are better than NH_3 treatment. Carbon coated iron impurities were liquified in an aqueous solution of H_2O_2 and HCl at 40-70°C for 4–8 h. The production of CNTs increased to approximately 50 wt. % and the purity raised up to 96 wt. % with this treatment [62].

4.3 Intercalation method

Halogen may be intercalated into carbon nanotubes for selective oxidation of carbonaceous impurities. Brominating is one of the effective procedures in CNTs purification process. Graphite intercalation compounds are formed by the attachment of atomic or molecular layers of a different chemical species between layers in graphite host materials. The intercalation of bromine (Br_2) in CNTs is confirmed by using HR-TEM [48]. The mixing of raw CNTs with pure liquid bromine under nitrogen atmosphere yielded Br. Under these conditions, charge transfer between Br and carbon occurred, enabling the formation of complex C-Br_2 on CNT surface and at deformed sites. It was observed that the orientation of Br on CNT surface is like a rod of a wheel, which is perpendicular to the curve of graphitic layers on CNTs. Intercalation of Br usually, happens on the surface of CNTs, where large numbers of defected sites are available. Br will be more reactive to those regions where different types of defects (amorphous carbon and other disorder carbonaceous materials) exist. When brominated, CNTs were passed through the air combustion at 550°C, and it was observed that the layers of the graphite were damaged along the line in which Br collected, showing the effect on the reactivity of the tubes toward oxygen upon adding Br. The amorphous carbon can be effectively oxidized due to the oxidation difference between brominated regions and CNTs. The catalyst particles, which was bounded, were opened and removed at the same time. Due to the tube action, Br diffused into the tubular CNTs and caused in the breakage of inner graphite layers during oxidation [63–65].

5. Conclusion

A detailed overview of synthesis and purification of carbon nanotubes is presented in this chapter. Synthesis techniques (i.e., arc discharge synthesis, laser ablation of graphite/laser vaporization synthesis method, chemical vapor deposition (CVD), high pressure carbon monoxide synthesis and flame synthesis) have been described in detail to highlight their importance as well as drawbacks. Arc discharge synthesis method is one of the most used technique for carbon nanotubes in large quantities. Its main drawback is the lack of control over the chirality in the nanotubes. Laser ablation method has the ability to produce CNTs in large quantities having small impurities. However, it is an expensive method as compare to arc discharge method for the synthesis of CNTs. A high purity CNTs can be obtained by using Chemical vapor deposition method. It is most suitable for large-scale manufacturing of CNTs at economical cost than laser ablation method. Chemical-based purification methods (i.e., gas phase, liquid phase and intercalation method) for CNTs are discussed comprehensively. These methods can efficiently eliminate amorphous carbon, polyhedral carbon and metal impurities at the cost of decreasing a significant amount of CNTs or damaging structure of CNTs. Gas phase purification is considered for purifying CNTs because it does not significantly grow sidewall defects in CNTs. However, it has limitation that it does not remove metal particles straightforwardly. Liquid phase oxidation produces defects on CNTs sidewall and may break-down CNTs into shorter ones with different lengths. The intercalation is best suitable for purifying CNTs without destroying their alignment. These features of synthesis and purification methods of CNTs will help researchers to select between these different methods according to their requirements.

Author details

Syed Awais Rouf[1*], Zahid Usman[1], Hafiz Tariq Masood[2], Abdul Mannan Majeed[1], Mudassira Sarwar[1] and Waseem Abbas[1]

1 Division of Science and Technology, Department of Physics, University of Education, Lahore, Pakistan

2 Department of Physics, University of Sahiwal, Sahiwal, Pakistan

*Address all correspondence to: awais.physicist@gmail.com

References

[1] Hennrich F, Chan C, Moore V, Rolandi M, and Connell M O', Carbon nanotubes properties and applications, Taylor & Francis Group, 2006.

[2] Kroto H W, Heath J R, Brien S C O', Curl R F, and Smalley R E, C60: buckminsterfullerene, Nature. 1985:318(6042):162-163.

[3] Schematic structure of SWNT; 2014.

[4] The transmission electron microscope (TEM) images of a MWNT; 2014.

[5] Iijima S, Ichihashi T: Single-shell carbon nanotubes of 1-nm diameter, Nature. 1993:363:603-605.

[6] Iijima S: Helical microtubules of graphitic carbon. Nature. 1991:354(6348):56-58.

[7] Overney G, Zhong W, Tomanek D, Structural rigidity and low frequency vibrational modes of long carbon tubules, Z. Phys. D 1993:27:93-96.

[8] Rinzler A G, Hafner J H, Nikolaev P, Lou L, Kim S G, Tomanek D, Nordlander P, Colbert D T, Smalley R E, Unraveling nanotubes: field emission from an atomic wire, Science. 1995:269:1550.

[9] Bachtold A, Hadley P, Nakanishi T, Dekker C, Logic circuits with carbon nanotube transistors, Science. 2001:294:1317.

[10] Che G, Lakshmi B B, Fisher E R, Martin C R, Carbon nanotubule membranes for electrochemical energy storage and production, Nature. 1998:393:346.

[11] Dillon A C, Jones K M, Bekkendahl T A, Kiang C H, Bethune D S, Heben M J, Storage of hydrogen in single-walled carbon nanotubes, Nature. 1997:386:377.

[12] Khlobystov A N, Britz D A, Wang J, O'Neil S A, Poliakoff M, and Briggs G A D, Low temperature assembly of fullerene arrays in single-walled carbon nanotubes using supercritical fluids, J. Mater. Chem. 2004:14(19): 2852-2857.

[13] Wongwiriyapan W, Honda S, Konishi H, Mizuta T, Ohmori T, Kishimoto Y, lto T, Maekawa T, Suzuki K, Ishikawa H, Murakami T, Kisoda K, Harima H, Oura K, and Katayama M, Influence of the growth morphology of single-walled carbon nanotubes on gas sensing performance, Nanotechnology. 2006:17(17):4424.

[14] Kong J, Franklin N R, Zhou C, Chapline M G, Peng S, Cho k, and Dai H, Nanotube molecular wires as chemical sensors, Science. 2000:287(5453): 622-625.

[15] Modi A, Koratkar N, Lass E, Wei B, and Ajayan P M, Miniaturized gas ionization sensors using carbon nanotubes, Nature. 2003:424(6945):171-174.

[16] Gao B, Kelinhammes A, Tang X P, Bower C, Wu Y, Zhou O, Electrochemical intercalation of single-walled carbon nanotubes with lithium, Chem. Phys. Lett. 1999:307:153.

[17] Ajayan P M, Zhou O Z. Applications of carbon nanotubes. In: Dresselhaus M S, Dresselhaus G, Avouris P, editors. Carbon Nanotubes. Berlin/ Heidelberg:Springer-Verlag; 2001. pp. 391-425.

[18] Baughman R H, Zakhidov AA, De Heer WA. Carbon nanotubes—The route toward applications. Science.2002;297:787-792.

[19] Ajayan P M, Ebbesen T W: Nanometre-size tubes of carbon. Rep. Prog. Phys. 1997:60(10):1025.

[20] Zhu H W, Xu C L, Wu D H, Wei B Q, Vajtai R, and Ajayan P M, Science. 2002:296(5569):884-886.

[21] Chico L, Crespi V H, Benedict L X, Louie S G, and Cohen M L: Pure carbon nanoscale devices: nanotube heterojunctions. Phys. Rev. Lett. 1996:76(6):971-974.

[22] Grobert N: Carbon nanotubes—becoming clean. Mater Today. 2007:10(1):28-35.

[23] Vander Wal R L, Berger G M, and Ticich T M, Carbon nanotube synthesis in a flame using laser ablation for in situ catalyst generation, Appl Phys A. 2003:77(7):885-889.

[24] Ganesh E N: Single Walled and Multi Walled Carbon Nanotube Structure. Synthesis and Applications 2013:2(4):311-318.

[25] Ferreira F V, Franceschi W, Menezes B R C, Biagioni A F, Coutinho A R, and Cividanes L S, synthesis, characterization, and applications of carbon nanotubes, in Carbon-Based Nanofillers and Their Rubber Nanocomposites, 2019.

[26] Dai H, Carbon nanotubes: opportunities and challenges, Surface Science, 2002:500(1-3):218-241.

[27] Bethune D S, Kiang C H, De Vries M S, Gormaan G, Savoy R, Vazquez J, and Beyers R, Cobalt- catalysed growth of carbon nanotubes with single-atomic-layer walls, Nature. 1993:363(6430):605-607.

[28] Askeland D R, Phul P P, The science and engineering of materials; 4th Edition, Brooks/Cole Publishing/ Thompson Learning, USA. 2003.

[29] Saito R, Dresselhaus G, Dresselhaus M S: Physical properties of carbon nanotubes. 4th edition. USA: World Scientific; 1998.

[30] Abbasi E, Sedigheh Fekri A, Abolfazl A, Morteza M, Hamid Tayefi N, Younes H, Kazem N K, Roghiyeh P A: Dendrimers: synthesis, applications, and properties. Nanoscale Research Letters 2014:9(1):247-255.

[31] Thess A, Lee R, Nikolaev P, Dai H, Petit P, Robert J, Xu C, Lee Y H, Kim S G, Rinzler A G: Crystalline ropes of metallic carbon nanotubes. Science-AAAS-Weekly Paper Edition, 1996:273(5274):483-487.

[32] Ncube S, Electronic properties of single walled carbon nanotubes synthesized by laser ablation, Materials Science, (2014).

[33] Vander Wal R L, Berger G M, Ticich T M: Carbon nanotube synthesis in a flame using laser ablation for in situ catalyst generation. Applied Physics A, 2003:77(7):885-889.

[34] Iijima S, Ajayan P M, Ichihashi T, Growth model for carbon nanotubes. Phys Rev Lett, 1992:69(21):3100.

[35] Journet C, Maser W K, Bernier P, Loiseau A, De La Chapelle M L, Lefrant D, Deniard P, Lee R, Fischer J E: Large-scale production of single-walled carbon nanotubes by the electric-arc technique. Nature, 1997:388(6644):756-758.

[36] He Z B, Maurice J L, Lee C S, Cojocaru C S, Pribat D: Nickel catalyst faceting in plasma-enhanced direct current chemical vapor deposition of carbon nanofibers. The Arabian Journal for Science and Engineering, 2010:35(1C):11-19.

[37] Dervishi E, Li Z, Xu Y, Saini V, Biris A R, Lupu D, Biris A S: Carbon nanotubes: synthesis, properties, and applications. Part Sci Technol 2009:27(2):107-125.

[38] https://sites.google.com/site/nanomodern/Home/CNT/syncnt/cvd

[39] Eatemadi A, Daraee H, Karimkhanloo H, Kouhi M, Zarghami N, Akbarzadeh A, Abasi M, Hanifepour Y, and Woo Joo S, Carbon nanotubes: properties, synthesis, purification, and medical applications. Nanoscale Res Lett. 2014:9:393.

[40] Park T J, Banerjee S, Hemraj-Benny T, Wong S S, Purification strategies and purity visualization techniques for single-walled carbon nanotubes, Journal of Materials Chemistry 2006:16(2):141-154.

[41] Haddon R C, Sippel J, Rinzler A G, Papadimitrakopoulos F, Purification and separation of carbon nanotubes, MRS Bulletin 2004, 29(4):252-241.

[42] Bonard, J M, Stora T, Salvetat J P, Maier F, Stöckli T, Duschl C, Forró L, de Heer W A, Châtelain A, Purification and size-selection of carbon nanotubes, Adv. Mater. 1997:9(10):827-831.

[43] Liu X M, Spencer J L, Kaiser A B, Arnold W M, Selective purification of multiwalled carbon nanotubes by dielectrophoresis within a large array, Curr. Appl. Phys. 2006:6(3):427-431.

[44] Ebbesen T W, Ajayan P M, Hiura H, Tanigaki K, Purification of nanotubes, Nature. 1994:367:519.

[45] Banerjee S, Wong S S, J. Phys. Chem. B 2002:106:12144.

[46] Dalton A B, Stephan C, Coleman J N, et al., Selective interaction of a semi-conjugated organic polymer with single-wall nanotubes, Journal of Physical Chemistry B 2000:104(43): 10012-10016.

[47] Coleman J N, Dalton A B, Curran S, et al., Phase separation of carbon nanotubes and turbostratic graphite using a functional organic polymer, Adv. Mater. 2000:12(2):213-216.

[48] Mahalingam P, Parasuram B, Maiyalagan T, Sundaram S, Chemical methods for purification of carbon nanotubes-A review, J. Environ. Nanotechnol. 2012:1(1):53-61.

[49] Scaccia S, Carewska M, Prosini P P, Thermochim. Acta. 2005:435:209.

[50] Smith M R, Hedges S W, Lacount R, Kern D, Shah N, Huffman G P, Bockrath B, Carbon. 2003:41:1221.

[51] Vivekchand S R C, Govindaraj A, Seikh M, Rao C N R, J. Phys. Chem. B. 2004:108:6935.

[52] Li C, Wang D, Liang T, Wang X, Wu J, Hu X, Liang J, Powder Technol. 2004:142:175.

[53] Ando Y, Zhao X, Shimoyama H, Carbon. 2001:39:569.

[54] Delpeux S, Szostak K, Frackowiak E, Beguin F, Chem. Phys. Lett. 2005:404:374.

[55] Ando Y, Zhao X, Shimoyama H, Carbon. 2001:39:569.

[56] Gajewski S, Maneck H E, Knoll U, Neubert D, Dorfel I, Mach R, Strauß B, Friedrich J F, Diamond Relat. Mater. 2003:12:816.

[57] Wiltshire J G, Khlobystov A N, Li L J, Lyapin S G, Briggs G A D, Nicholas R J, Chem. Phys. Lett. 2004:386:239.

[58] Cheap Tubes Inc. Purification techniques for carbon nanotubes including gas and liquid phase and intercalation. AZoNano. 17 April 2021. https://www.azonano.com/article.aspx?ArticleID=1562

[59] Yoshiro O, Oxidations with Nitric Acid or Nitrogen Oxides. 1978: 295-342.

[60] Raymundo-Piñero E, Cacciaguerra T, Simon P, Béguin F, A single step process for the simultaneous purification and opening of multiwalled

carbon nanotubes, Chem. Phys. Lett. 2005:412(1-3):184-189.

[61] Joncourt L, Mermoux M, Touzain P H, Bonnetain L, Dumas D, Allard B, Sodium reactivity with carbons, J. Phys. Chem. Solids, 1996:57(6-8):877-882.

[62] Hammadi A H, Jasim A M, Abdulrazzak F H, Al-Sammarraie A M A, Cherifi Y, Boukherroub R, Hussein F H, Purification for carbon nanotubes synthesized by flame fragments deposition via hydrogen peroxide and acetone, Mater. 2020:13:2342.

[63] Hamwi A, Alvergnat H, Bonnamy S, Beguin F, Carbon. 1997:35:723.

[64] Hou P X, Bai S, Yang Q H, Liu C, Cheng H M, Carbon. 2002:40:81.

[65] Jin Z X, Xu G Q, Goh S H, Carbon. 2000:38:1135.

Section 3

Natural Rubber Carbon Nanotube Composites for Electronics Applications

Chapter 4

Carbon Nanotubes Reinforced Natural Rubber Composites

Apinya Krainoi, Jobish Johns, Ekwipoo Kalkornsurapranee and Yeampon Nakaramontri

Abstract

Several advanced methods have been introduced to disperse CNTs in the NR matrix. Various aspects highlighted in this chapter include the mixing processes such as melt mixing and latex mixing methods. As well as, formations of functional groups on the surfaces of CNT using silane coupling agents (i.e., ex-situ and in-situ functionalization). Moreover, hybrid CNT are beneficial to achieve better electrical conductivity of NR/CNT composites. These efforts are aimed to reduce the percolation threshold concentration in the NR composites for application as conducting composites based on electrically insulating rubber matrix. Sensor application is developed based on conducting NR composites. NR composites showed changing of resistivity during elongation termed as piezoresistivity. The most commonly used rubber matrices such as NR, ENR and IR are mixed with a combination of CNT and CB fillers as hybrid filler. The presence of linkages in the ENR composites results in the least loss of conductivity during external strain. It is found that the conductivity becomes stable after 3000 cycles. This is found to be similar to the NR-CNT/CB composite, while a few cycles are needed for IR-CNT/CB owing to the higher filler agglomeration and poor filler-rubber interactions. This is attributed to the polar chemical interactions between ENR and the functional groups on the surfaces of CNT/CB.

Keywords: natural rubber, carbon nanotube, nanocomposites, electrical conductivity, sensor

1. Introduction

Natural rubber (NR) is widely used in various industries owing to its excellent elasticity and mechanical properties. NR has been typically used in many industrial applications including tires, sports articles, sealing materials, medical glove, rubber boots and dairy rubber items [1]. Moreover, application of NR can be more applied by addition of fillers, such as silica, clay, carbon black, and carbon nanotube that is properties of NR can be induced by filler. As NR was converted from insulator material to be used as semi-conductive.

CNT have been widely interested for using as conductive filler in NR composites, due to the sp^2-hybridized carbon molecules throughout its molecular structure. Its carbon–carbon bond angles can be mechanically distorted reversibly, and core electrons can act as free electrons of the carbon atoms on CNT surfaces. Thus, the special molecular structure of CNT provides it with high mechanical properties, excellent thermal conductivity, and outstanding electrical conductivity [2].

Furthermore, nanocomposite of NR and CNT provided high elasticity material and also sensitivity on electricity due to CNT networks in NR matrix are easy to break under stretching and fast rebuilt under releasing [2]. Therefore, it well proper for application as smart sensors to monitor the applied external stimulus. That is, NR/CNT composites based stretchable strain sensors have been interested to emerging applications, such as human motion detection [3, 4].

However, several works have been researched on human motion detection as adsorbing graphene woven fabrics on polydimethylsiloxane (PDMS) and medical tape composite. The wearable strain sensor could well detect human movements, including hand clenching, pulse, expression change, blink, phonation, and breathing [4]. Additionally, it is observed that the stretchable CNTs/carbon black (CB)/ isoprene rubber (IR) composites could be used to detect human motions and emotional expressions [5]. It was reported that the percolation threshold concentration of composites was significantly increased, while optimal conductivity increased, on adding conductive CB in CNT composites [5]. Furthermore, using CB also improves the sensitivity of electrical resistivity to stress and strain, due to its spherical shape that eases disconnection of conductive particles by strain, while the long cylindrical CNT particles can have sliding contact. This increases potentially the piezoresistive responsiveness, combining excellent conductivity of CNT with strain sensitivity of the electrical pathways on using CNT-CB blended filler [6, 7]. Furthermore, using NR, incorporation of CNT and CB hybrid filler can keep a very stable sensor performance, showing good mechanical properties, when the composites are dynamically elongated several times [6, 8]. Also, NR composites are easy to process, cost-effective, and well-known as hydrophobic biopolymers [9], so that humidity does not effect on an NR sensor.

This review article focuses on the preparation and electrical property of NR/ CNT composites, the methods to improve the dispersion CNT are also mentioned as well as overview of applying NR/CNT composites for motions sensors application.

2. Properties of CNTs

Usually, CNT has extremely high tensile strength compared to other carbon materials. The excellent strength makes CNTs suitable for developing composite material with higher reinforcing efficiency. It was also found that, the incorporation of 0.5 phr MWCNT in NR composite reflected the best properties of increasing 61% of tensile strength and 75% of modulus [10]. Moreover, CNT exhibits excellent electronic properties as the details given in **Table 1** [11, 12].

Properties	SWCNT	MWCNT
Young's modulus (GPa)	100–500	20–95
Tensile strength (GPa)	15–53	11–63
Electrical conductivity (S/cm)	10^2–10^6	10^3–10^5
Electron mobility (cm^2/Vs)	~10^5	10^4–10^5
Thermal conductivity (W/m K)	6000	2000

Table 1.
Properties of carbon nanotube.

3. CNTs-based natural rubber composites

3.1 Modified natural rubber-carbon nanotube composites and its properties

Natural rubber (NR) is a well-known biopolymer that consists of isoprene units linked together in *cis*-1,4 configuration. NR has attracted tremendous scientific and industrial interests due to its unique molecular structure with superior and unique properties such as high elasticity, flexibility and some level of biodegradability. However, NR has intrinsically poor aging, weathering, oil resistance, and electrical conductivity that limits the use of NR in some applications. However, the application of NR can be extended by the modification of NR molecules in various forms, such as epoxidized natural rubber (ENR) and maleated natural rubber (MNR). Various properties of NR products (i.e., modulus, viscosity and strength) could be improved by incorporating special types of fillers to form NR composites. Therefore, different types of fillers into NR as an elastomeric matrix, including carbon black, silica, clay, calcium carbonate, carbon fibers or carbon nanotubes, have been widely investigated. CNT filled NR composites were used with various types of natural rubbers, especially unmodified natural rubber (NR), epoxidized natural rubber (ENR) and maleated natural rubber (MNR) [13]. Consequently, after the modification of rubber, the electrical conductivity of composites was found to be enhanced when compared to the unmodified NR-CNT (**Figure 1**). The percolation limit for CNTs in ENR-CNT and MNR-CNT composites is approximately 1 phr, while a value of 4 phr was found for unmodified NR-CNT composite. Lower percolation value of ENR-CNT and MNR-CNT composites than that of the unmodified NR-CNT composite proves the enhanced degree of CNT dispersion in the rubber matrix. It is gained by the occurrence of chemical interaction between the functional groups present in ENR or MNR molecules and the polar groups on CNT surfaces as shown in **Figure 2**. This confirms that, the polar nature of rubber molecules (i.e., ENR and MNR) causes significantly a greater degree of CNT dispersion and consequently the composites reach the percolation limit at smaller CNT concentration. However, for all the composites studied, the maximum electrical conductivities are the same. It means at the CNT concentrations

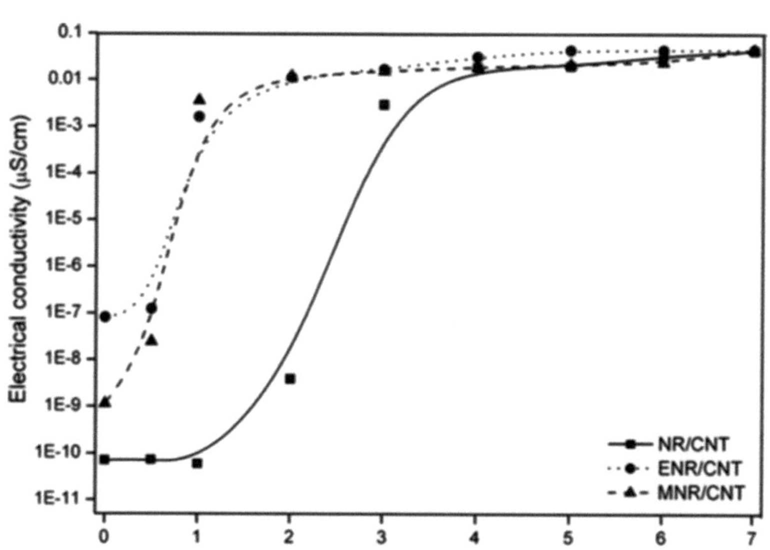

Figure 1.
Electrical conductivity of CNT-filled rubber composites with various CNT contents of 0–7 phr [2].

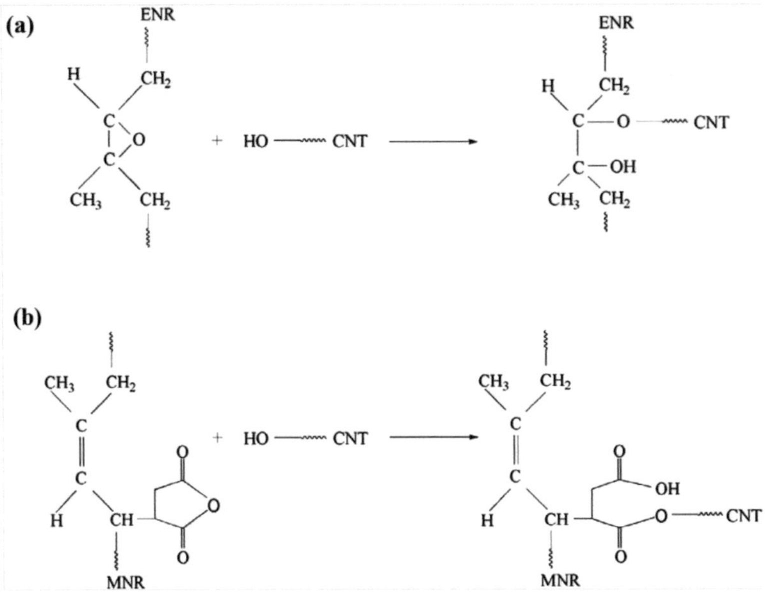

Figure 2.
Possible chemical reactions between (A) ENR and CNT, and (B) MNR and CNT [13].

below the percolation threshold the electrical conductivity is dominated by the rubber matrix. Whereas the CNT network plays the dominating role above the percolation limits. This correlates well to Earp *et al.* which indicated a comparable conductivity of the NR composites with and without CNT loading since the CNT was covered by the insulated NR molecules. After the increase of CNT concentration, the percolative CNT are formed and the NR matrix had CNT pathway dispersing throughout the samples. This CNT connection induces and carries well transferring of electron which causes increment of electrical conductivity [14]. This is agreement to Ma *et al.* which found that electrically conducting behavior of composites consisting of conducting fillers and insulating matrices can be applied to explain the percolation theory originating in the materials. It was found that the composite undergoes an insulator-to-conductor transition while the conducting filler content is gradually increased. The critical filler concentration is referred to the percolation threshold where the measured electrical conductivity of the composite had sharply increased for several orders of magnitude relating the formation of continuous electron paths or conducting networks [11]. Moreover, critical CNT loading in matrix effects on the overall properties of CNT filled ENR nanocomposites [2]. On varying the CNT loading from 1 to 7 phr showed the critical loading at 3 phr and significantly improved the electrical conductivity.

3.2 Dispersion technique of carbon nanotubes and their network formations on the properties of natural rubber-carbon nanotube composites

Recently, CNT becomes a promising filler for the NR based composites due to its several unique properties. Perfect molecular structure of CNT with sp^2-hybrided carbon structure causes extremely high mechanical properties, excellent thermal conductivity and outstanding electrical conductivity [11]. In addition, low density, high specific surface area and extremely high aspect ratio make the CNT as an interesting carbon filler same as graphene and other carbon fibers. In the recent years, many researchers have attempted to incorporate CNT into rubber matrix (i.e., natural rubber [15–17] and synthetic rubber [18, 19]) to utilize the intrinsic properties of CNT for enhancing the properties of rubber composites, particularly for the electrical conductivity. However, the property enhancement is not so easy and still the vigorous investigations are ongoing. The major drawback to use CNTs as the reinforcing filler in NR is their agglomeration, since CNT contains very high aspect ratio and strong Van-der Waals attractions between the particles. Small polar functional groups on the CNT surface are also the reason for their self-association behavior inside NR matrix. Altogether it provides strong filler-filler interaction which causes very poor dispersion of CNTs. Weak physical and chemical interactions among CNT and NR matrix generally lead to poor mechanical properties and electrical conductivity due to the incompatibility between them [20]. Therefore, homogeneous dispersion of CNTs inside the rubber matrix is an important challenge by optimizing the condition for the preparation of rubber-CNT composites. To obtain high conductive CNT-based rubber composites, a proper preparation method has also been widely investigated. Melt blending and latex state mixing processes are the most effective methods in terms of the process ability and properties of nanocomposites by using a two-roll mill and an internal mixer [21]. Shearing force and mixing temperature during rubber operation cause reduction of NR viscosity and therefore the CNT can be easily dispersed and distributed in NR matrix. However, this mixing system had originated much the heat and not environmentally friendly operation owing to dispersion of low density of CNT. Thus, latex-based composites are represented and it showed significantly improved properties than relative to the one preparing from melt mixing. It was found that the lowest percolation threshold concentration of approximately 0.5 phr of CNTs was observed in the latex–CNT composites [22]. Electrical conductivity is one of the properties that can be applied to characterize the quality of filler dispersion in CNT composites. If a continuous filler network of electrically conductive fillers is formed, the material undergoes a sudden transition from insulator to conductor. As a result, the electrical conductivity rises by several orders of magnitude. **Figure 3** shows the effect of filler loading on the electrical conductivity of CNT-filled composites based on NR from ADS and latex. Here, the latex-based composites exhibited a percolation threshold at a CNT concentration lower than 1 phr. This is due to the orientation of nanotubes along a specific path around the rubber particles which resulted in the formation of segregated nanotube network [23, 24] as confirmed by the TEM image (**Figure 4**).

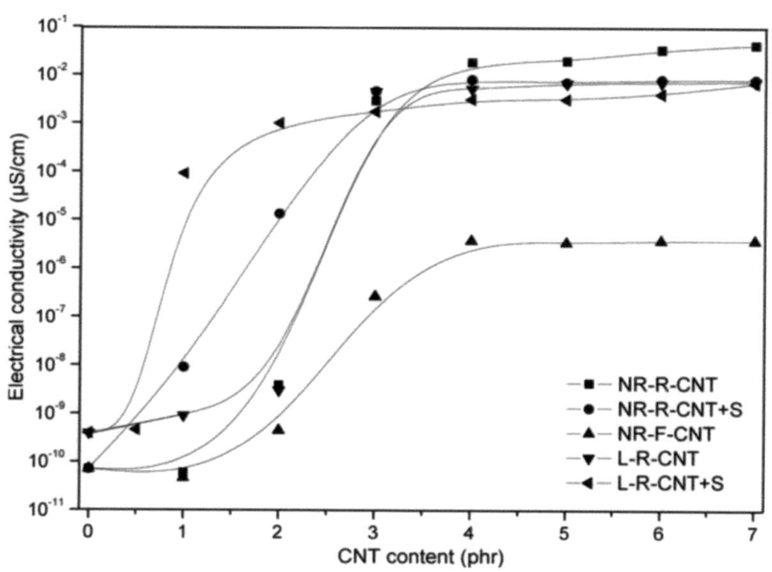

Figure 3.
Electrical conductivity of composites as a function of CNT content [22].

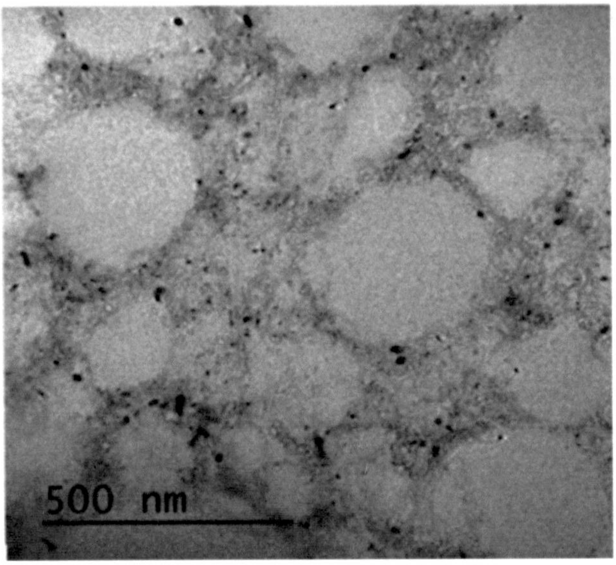

Figure 4.
TEM image of natural rubber composite film containing segregated network of MWCNT [23].

3.3 Functionalization of carbon nanotubes and the properties of natural rubber-carbon nanotube composites

The major drawback of CNT as a reinforcing filler in NR is its agglomeration tendency, since the CNT fibers have very high aspect ratio and strong Van-der Waals attraction between each other. This is due to the lack of polar functional groups on the CNT surfaces which also leads to the self-association behavior in the NR matrix. Generally, the filler-filler interactions are too strong compared to filler-matrix interactions causes very poor dispersion of the filler. The poor physical and chemical bonding between CNT and NR or the incompatible nature generally leads to exhibit poor stability of the composites in terms of their mechanical properties and electrical conductivity [20]. Therefore, attaining a homogeneous dispersion of CNTs in the rubber matrix remains a challenge and addressed by seeking the optimal conditions for the preparation of rubber-CNT composites. To improve the dispersion of CNTs in the NR matrix, a silane coupling agent was applied by expecting that the filler-rubber interactions would be enhanced by reducing the Van-der Waals attractions of CNT particles. The *ex-situ* functionalization of CNTs with silane has been introduced to improve the CNT dispersion in rubber-CNT composites. However, this method is time consuming and more expensive, and might not be appropriate in practical applications. Thus, recent studies have investigated *in-situ* functionalization of CNTs with silane. Similar to rubber-silica composites [25, 26], silane was added directly during the mixing of rubber and silica. The silanization of silica particles can take place during mixing if the mixing conditions are suitable. This alternative process can provide a short processing cycle compared to *ex-situ* silanization. On the other hand, the functional groups on the raw CNTs are readily available and sufficient to react with silane molecules during mixing [22] as similar to the silica-filled composites. Thus, the way of mixing is most important to improve the reinforcing efficiency of CNTs in rubber-CNT composites [27, 28]. CNT filled NR composites were prepared by melt mixing and latex mixing methods. The *in-situ* functionalization of CNTs with a silane coupling agent, namely bis (triethoxysilylpropyl) tetrasulfide (TESPT) was done to improve

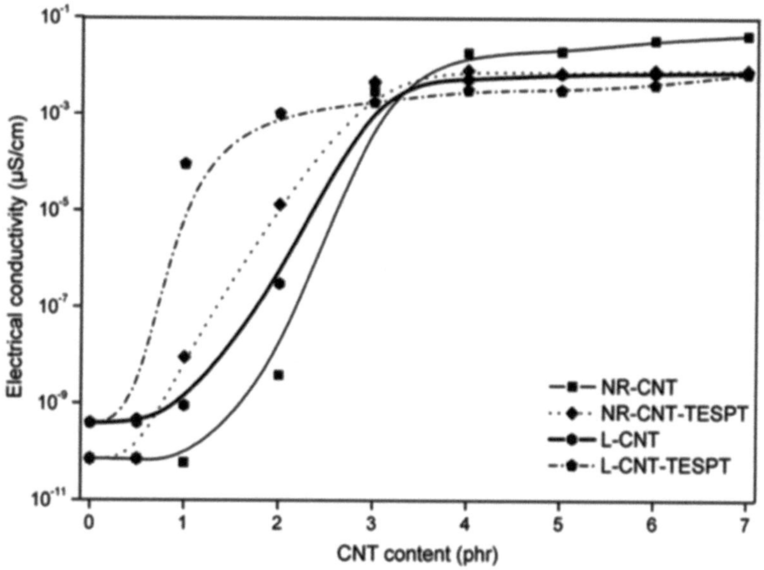

Figure 5.
Variation of electrical conductivity as a function of CNT content for various NR–CNT composites [30].

the interactions between CNT surfaces and rubber molecules. **Figure 5** shows the effect of CNT loading on the electrical conductivity of rubber–CNT composites with and without TESPT prepared by melt mixing and latex mixing processes. The lowest percolation threshold was observed in the composites prepared by latex mixing with *in-situ* functionalization. This is due to the chemical interactions of CNT surfaces, silane, and NR molecules (**Figure 6**) that improved the CNT dispersion and reduced the electrical percolation threshold. As a result, percolation thresholds were observed at approximately 2 and 1 phr of CNTs in NR–CNT–TESPT and L–CNT–TESPT composites, respectively. In addition, it was obtained the same trend as NR and ENR vulcanizates reinforced with CNT, CCB and CNT/CCB hybrid filler that decreasing of physically bound rubber absorption with addition of TESPT are showed, while the chemically bound amount had significantly increased. It was also found that superior conductive material with low dielectric constant of NR and ENR vulcanizates with CNT and CCB hybrid filler are received after the addition of TESPT [29].

In addition, composites of CNT and ENR were also prepared with in-situ functionalization of CNT with two alternative silane coupling agents such as bis(triethoxysilylpropyl) tetrasulfide (TESPT) and 3-aminopropyltriethoxysilane (APTES). The reactions of ENR molecules with the functional groups present on the CNT surfaces and also with the silane molecules were schematically shown in **Figures** 7 and **8**.

Composites of ENR–CNT and ENR–CNT–TESPT were successfully prepared with a very low electrical percolation threshold at 1 phr CNT content as showed in **Figure 9**. Furthermore, the highest electrical conductivity was achieved in the ENR–CNT–TESPT composite, due to its higher cross-link density and near-optimal CNT dispersion. Moreover, the morphological study of ENR–CNT and ENR–CNT–TESPT composites was used to confirm the fine dispersion of CNTs in the ENR matrix with loosely agglomerated CNTs. Consequently, the composites of ENR–CNT and ENR–CNT–TESPT exhibited improved tensile properties with higher cross-link density and electrical conductivity than the baseline of pristine ENR.

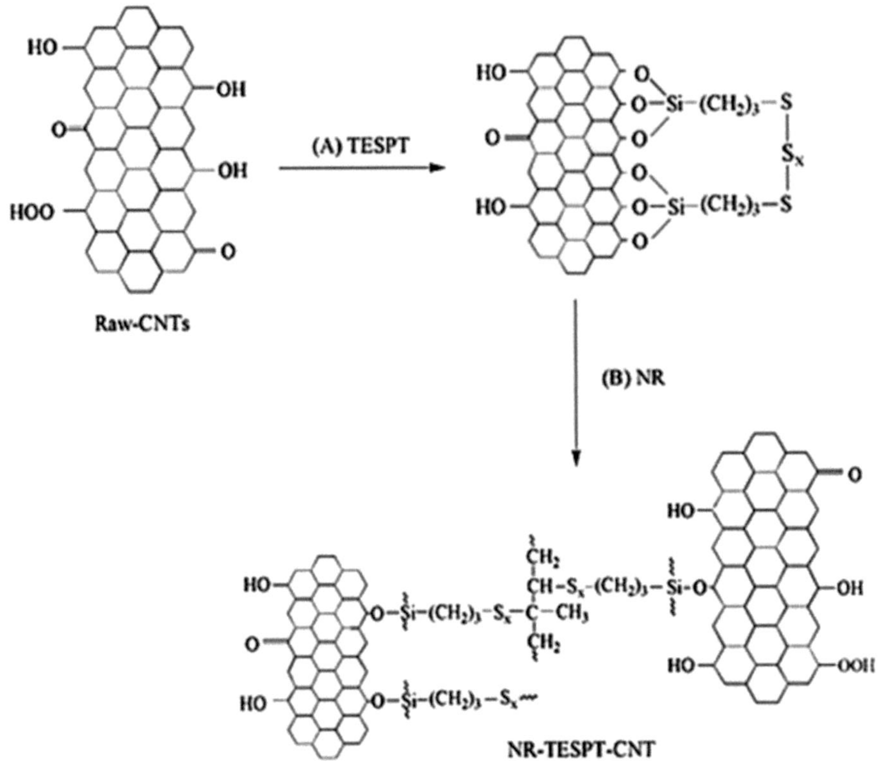

Figure 6.
Chemical reactions of functional groups on CNT surfaces and TESPT molecules (A), and silanized CNTs and NR molecules (B) [30].

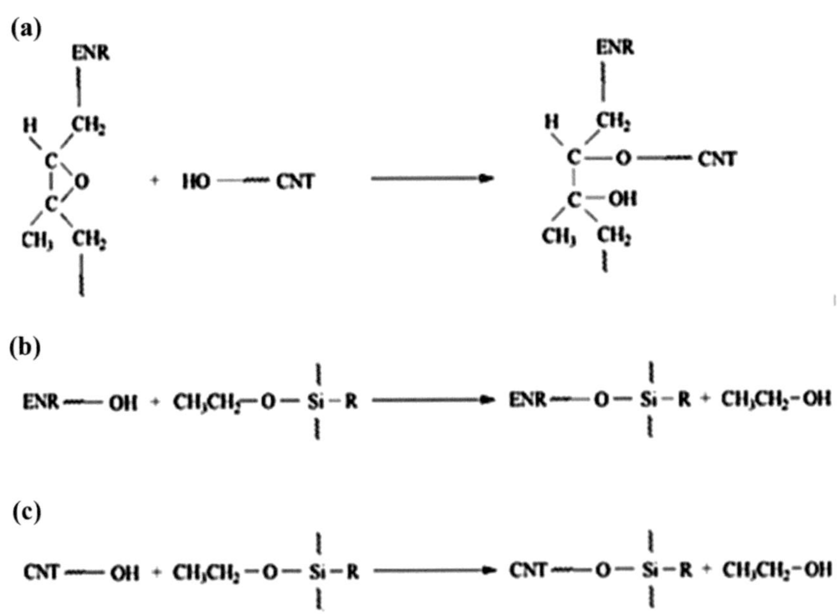

Figure 7.
Possible chemical reactions among (a) oxirane ring of ENR and functional groups of CNT, (b) hydroxyl groups of ENR and ethoxy groups of silanes, and (c) hydroxyl groups of CNT and ethoxy groups of silane molecules [9].

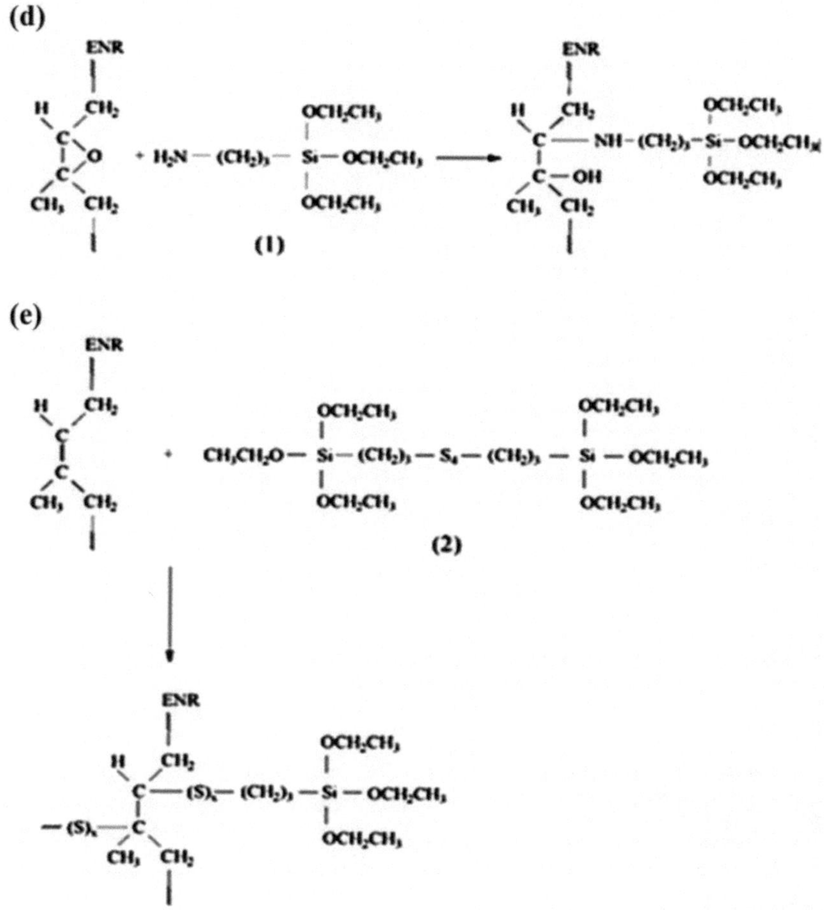

Figure 8.
Possible chemical reactions among (d) oxirane ring of ENR and amino group of APTES and (e) double bond of ENR and sulfur of TESPT molecules, where (1) and (2) are the molecular structures of APTES and TESPT [9].

3.4 Hybrid carbon nanotubes filled natural rubber composites

Several attempts have been made to disperse the CNTs in NR matrix by avoiding its re-agglomeration. To overcome this limitation, the addition of secondary fillers was introduced into the composites by generating new conductive hybrid filler pathways [32, 33]. An improved conductivity was achieved by adding carbon black (CB) into the CNT polymer composites [34–37]. Electrical conductivity of the composites was found to be slightly increased with CB concentration when the CNT content lies below its percolation threshold. However, no significant increase in the electrical conductivity occurred above the percolation threshold concentration. This might be due to the agglomeration of CB connected to CNT surfaces, which impedes the conductivity of hybrid ternary composites [35]. Thus, the CB can bridge CNT encapsulates and contribute new electron pathways only with highly homogeneous distribution of both the fillers. In this regard, the extremely high viscosity of NR is essential to enhance the conductivity by enabling good dispersion of fillers during mixing. No prior studies have been reported on the NR vulcanizates to assess the electrical conductivity with the dual fillers CB and CNT. A hybrid epoxy-based nanocomposite was developed by reinforcing CNT and CB. It was found that, the gaps between carbon nanotubes were

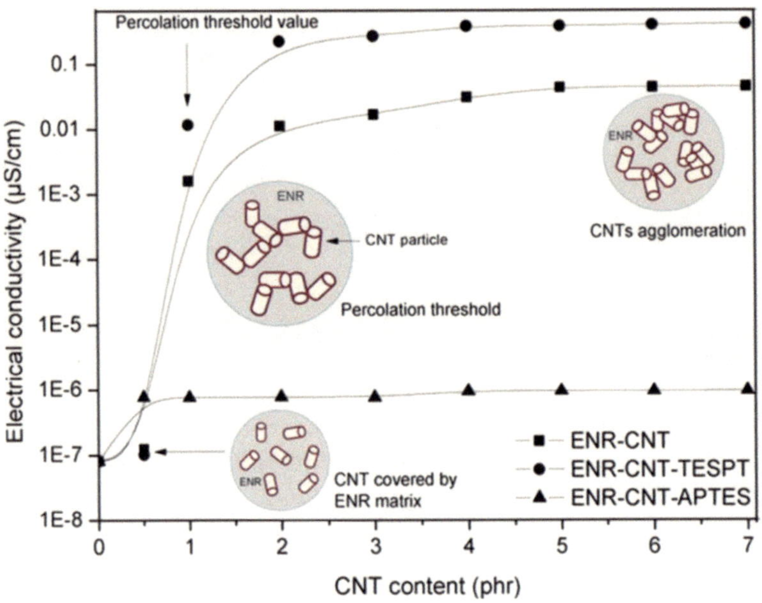

Figure 9.
Electrical conductivity of ENR–CNT, ENR–CNT–TESPT, and ENR–CNT–APTES composites prepared by in-situ functionalization with various CNTs contents of 0–7 phr [31].

connected by the CB nanoparticles, causing the formation of conducting networks [32, 34] as shown in **Figure 10**. The same behavior was observed in the hybrid of expanded graphite (EG)-CNT filled cyanate ester (CE) [38], graphene nanoplatelets (GNPs)-CNT/epoxy composites, titania nanoparticles (TiO_2)-CNT/epoxy composites [39] and hybrid of Ag nanoparticles (Ag-NPs)-CNT [40].

3.4.1 Hybrid composites of carbon nanotubes and conductive carbon black reinforced natural rubber

Filled NR vulcanizates were prepared by incorporating carbon-based fillers, namely carbon nanotubes (CNT), conductive carbon black (CCB), and CNT/CCB hybrid filler [41]. Reinforcement of CNT and CCB was carefully done by using a two-roll mill. The main aim was to generate an optimal state of filler dispersion in the NR matrix, in which CCB particles/aggregates bridge the CNT encapsulates. It improved the optimum electrical conductivity of NR composites by enabling electron tunneling and it is appropriate to provide fillers in the NR matrix. It was expected that, the achievable conductivity would synergistically be better than those of rubber composites with solely CNT or CCB. The variation of conductivity (at $f = 1$ Hz) with the filler volume fraction according to the Voet model is shown in **Figure 11**. It is seen that, the increment in conductivity appears in different steps for the NR vulcanizates filled with CCB (4 steps), CNT (3 steps), and CNT/CCB hybrid filler (2 steps). As already stated, there is no percolation threshold observed in case of CCB filler in NR vulcanizate, even though increasing the CCB loading up to 15 phr. In **Figure 11**, the NR vulcanizate filled with CNT/CCB hybrid filler showed only two step increments in conductivity. It is also clear that the filled NR vulcanizate had linear conductivity in between 1to10 phr of CCB in the CNT/CCB hybrid filler and saturates at 15 phr of CCB owing to the strong agglomeration. This means that

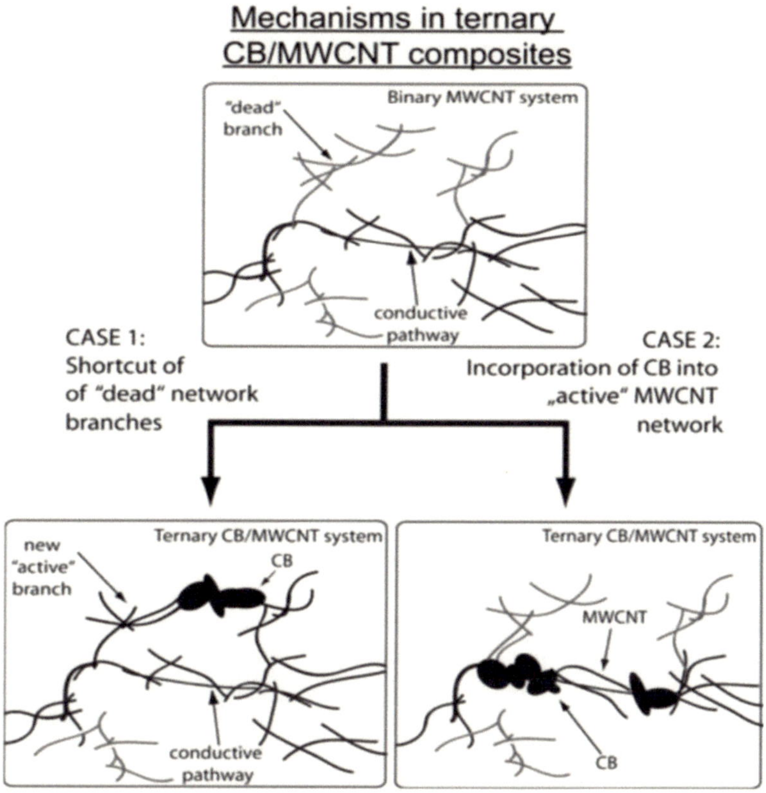

Figure 10.
Principles of conductive pathway formation in ternary CB/MWCNT systems [32].

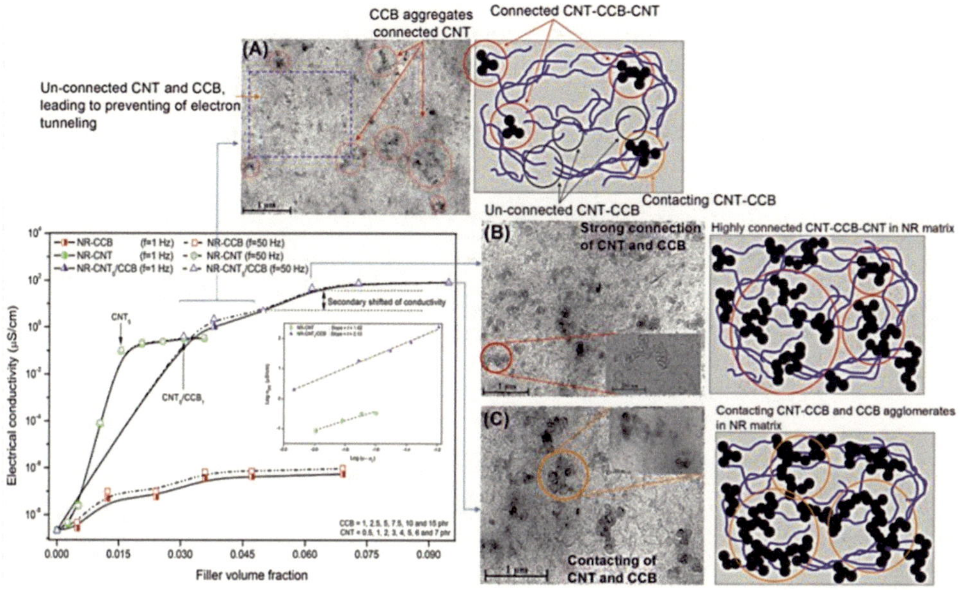

Figure 11.
Electrical conductivity of the NR vulcanizates filled with CCB, CNT, and CNT/CCB hybrid filler at various filler loadings [41].

NR vulcanizate filled with 5 phr CNT is an Ohmic conductor in between 1to10 phr of added CCB. It confirms the synergistic effect of CNT and CCB fillers in the NR vulcanizates, that improved and extended the conductivity of the NR composites by enhancing the electron tunneling and reducing the gaps between CNT encapsulates.

3.4.2 Hybrid carbon nanotubes and silver nanoparticle in natural rubber composites

The conductive NR composite with CNT-decorated AgNP (**Figure 12**) was prepared via the latex mixing method to get homogenous dispersion of the filler [42]. The decoration of CNT surfaces with AgNP significantly enhanced the electrical conductivity and lowered the percolation threshold concentration of NR composites when compared to the composites with plain CNT filler.

The percolation threshold concentrations of CNT and CNT-AgNP filled NR composites (**Figure 13**) are found to be 3.64 and 2.92 phr respectively. The combination of AgNP with CNT hybrid filler caused decreasing the percolation concentration and significantly increasing the optimal conductivity of the NR composites. This is due to the network formation of CNT-AgNP in the NR matrix favors the flow of electrons as compared to the NR filled with solely CNT. Therefore, better movement of the electrons by tunneling throughout the NR matrix was encountered. The degree of network formation of fillers in rubber matrix can be estimated by the *t* values. In **Figure 13(b)** and **(c)**, the *t* values of CNT and CNT-AgNP filled NR vulcanizates are noticed as 2.34 and 1.86, respectively. This clarifies that the CNT-AgNP filled NR vulcanizates are fully three-dimensional networks of fillers in the NR matrix, whereas the CNT filled NR vulcanizates showed stronger CNT agglomeration as indicated by higher *t* value. It also confirms the bridging of AgNP with end-to-end of CNT in the NR matrix which usually improves significantly the electrical conductivity and the percolation threshold of the composite.

3.4.3 Hybrid carbon nanotubes and ionic liquid in natural rubber composites

To enhance the electrical conductivity of the rubber composites, several methodologies have been exploited by improving the CNT dispersion in rubber

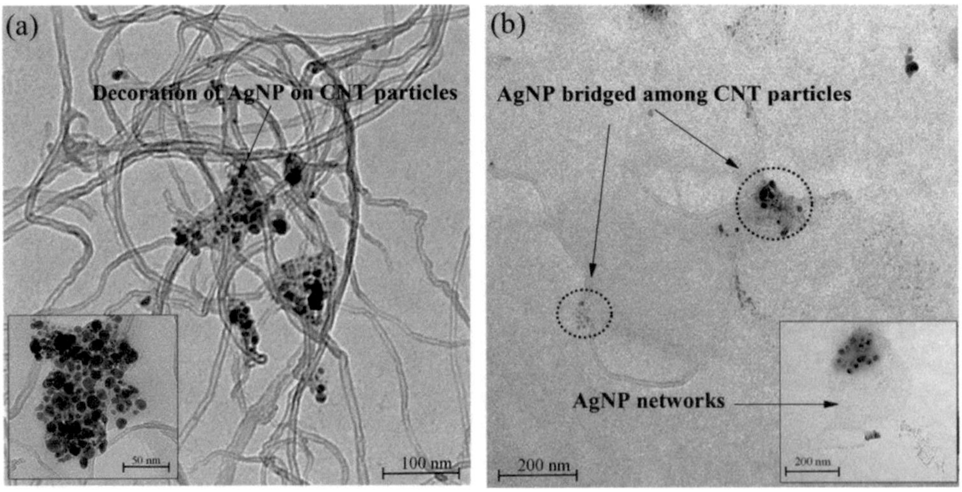

Figure 12.
Transmission electron microscopy (TEM) images of CNT decorated with silver nanoparticle (CNT-AgNP) [42].

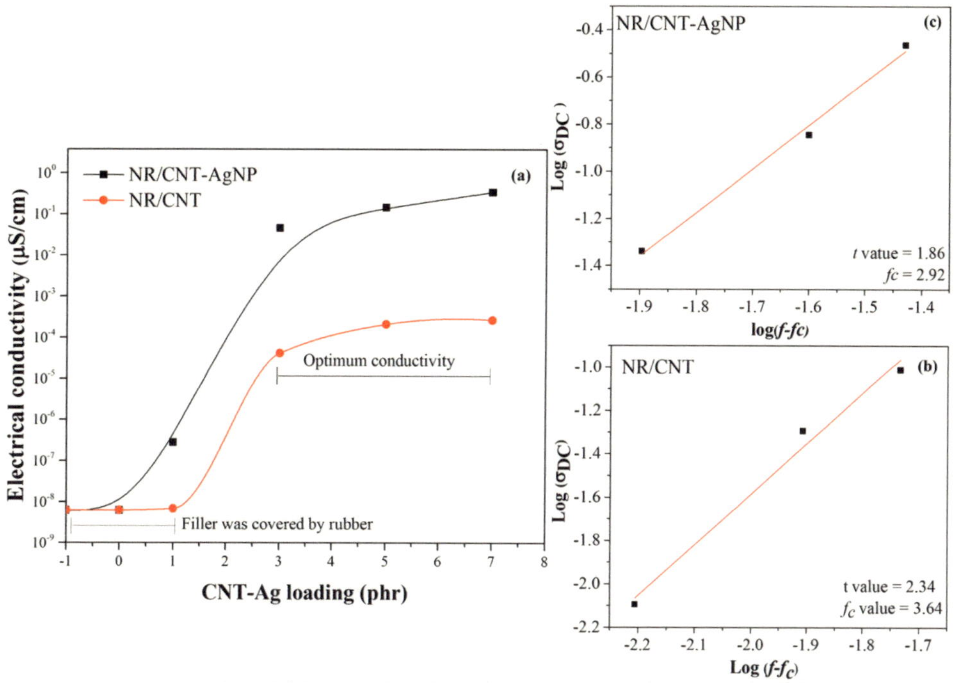

Figure 13.
Electrical conductivity of CNT and CNT-AgNP filled NR vulcanizates with various CNT and CNT-AgNP loadings (a), Plots of log σ_{dc} and log $(f\text{-}f_c)$ of CNT filled NR vulcanizates (b) and CNT-Ag filled NR vulcanizates (c) [42].

matrix. One prominent approach is the use of CNT mixed with an ionic liquid (IL) [43]. Typically, the IL molecules have hydrophilic and hydrophobic parts of the inorganic and organic salts in their molecules. It is noted that the hydrophobic part has the ability of interacting with CNT surfaces through cation- ð interaction [44]. Also, some ionic liquids contain -C=C- in the alkyl chain, and this could interact with diene rubbers *via* sulfur bridges in case of sulfur vulcanization system [45]. Therefore, IL forms bridge CNT surfaces with the rubber matrix [45]. The imidazolium ionic liquid has been widely used in various types of polymer matrix [46–49]. It was found that the imidazolium groups play an important role in improving the ionic conductivity of acrylonitrile butadiene rubber (NBR) [48]. Furthermore, NBR/SiO_2 in combination with imidazolium ionic liquid exhibited good elastomeric properties, high tensile strength, and high electrical conductivity [49]. In addition, CNT filled NR composites improved the conductivity by the addition of an ionic liquid (IL) 1-butyl-3-methyl imidazolium bis (trifluoromethylsulphonyl)mide (BMI) [50]. **Figure 14** clearly shows the addition of IL in to NR slightly increased the electrical conductivity, but the loading level of IL (BMI) does not significantly affect the conductivity of NR vulcanizate. This might be attributed to the encapsulated IL (BMI) by the insulating NR as the imidazolium IL could be more compatible with the hydrophobic rubber matrix [46]. This leads to reduce the electrical conductivity of the NR/IL vulcanizate with no noticeable percolation threshold. On the other hand, the composites of NR/CNT and NR/CNT-IL showed percolation threshold concentrations at 3.64 and 2.92 phr, respectively. Therefore, the NR/CNT-IL composite exhibited comparatively higher electrical conductivity with lower percolation threshold than the composite of NR/CNT. This might be due to the synergy of plasticizing by IL (BMI), contributed to good dispersion of CNT. It forms three-dimensional networks in the NR matrix

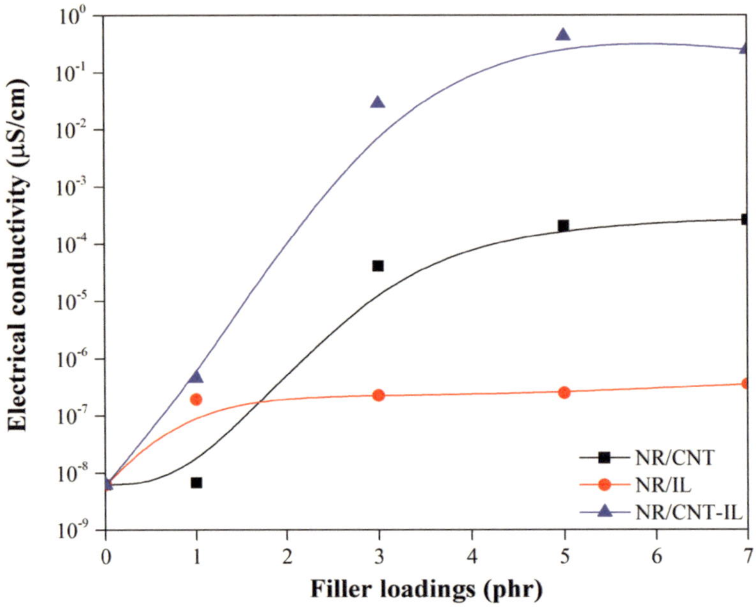

Figure 14.
Electrical conductivity of NR/CNT, NR/IL and NR/CNT-IL vulcanizates with various filler loadings [50].

assisted by the physical interactions of CNT particles. Therefore, the plasticizing effect and physical interactions facilitated CNT network formation and reduced the agglomeration of CNT.

4. Piezoresistive carbon-based composites for sensor applications

Conductive composites based on electrically insulating rubber matrix have attracted both scientific research and industrial interest for several years [11]. The two main parts in such composites are (i) the insulating rubber matrix and (ii) the conducting filler. The filler needs to form conductive pathways in the matrix for carrying electrons, thereby making the composite a semiconductor or a conductor [19]. Such filler pathways are perturbed by breakage and re-arrangement inside the matrix during deformation [33]. This change in resistivity during elongation is known as piezoresistivity and it can be used in motion detector applications [51]. Hence, the sensitivity of a composite sensor is affected by the type of rubber matrix and the type of fillers such as carbon black (CB), carbon fibre, graphene, graphite, and carbon nanotubes (CNT) [34, 35, 52, 53]. The CNT filled composites can serve in sensor applications due to its excellent electrical conductivity, which responds to various external stimuli such as temperature, organic solvents, vapour, strain, and damage [6]. Incorporation of CNT and CB hybrid filler in NR exhibits a very stable sensor performance along with good mechanical properties when the composites are dynamically elongated several times [6, 8]. Therefore, three alternative rubber matrices such as NR, epoxidized-NR (ENR) and isoprene rubber (IR) have been tested to clarify the effectiveness of the rubber matrix in a strain sensor containing CNT and CB as a hybrid filler. An appropriate ratio of CNT:CB was fixed at 1:1.5to form the filler networks throughout the matrix. Melt blending was selected as the mixing method to prepare the composites with the help of an internal mixer and a two-roll mill by optimizing the state of dispersion of fillers in the rubber matrix. Furthermore, the piezoresistivity (strain sensitivity of electrical resistance) was investigated in terms of the relative change in resistance, $\Delta R/R_0$ (ΔR is the change in resistance with strain, and R_0 is the initial resistance of the composite) [6, 54]. The measurement was performed with the help of an instrumental setup as showed in **Figure 15**.

To assess the effects of long term deformations on the composites, dynamic cyclic tensile testing at 50% strain for 50, 100, 500, 1000, 3000, 5000 and 10000 cycles was performed with an extension speed of 200 mm/min. Here, the resistance of the composites after each run was noticed. **Figure 16** shows the electrical conductivity as a function of cycle count for NR, ENR and IR composites with CNT/CB hybrid filler. The conductivity of these composites was found to be decreased with cycle count. The linkages in ENR composites exhibited the least loss of conductivity. It was found that the conductivity becomes stable after 3000 cycles (from 15.4 μS/cm to 0.044 μS/cm at 3,000 rounds). This is similar to the composites of NR-CNT/CB, while a few cycles were needed for IR-CNT/CB owing to the higher filler agglomeration and poor filler-rubber interactions. This is attributed to the polar chemical interactions between ENR and the functional groups on the surfaces of CNT/CB. Furthermore, the non-rubber components in NR and ENR matrices improved the filler dispersion as seen in the TEM images of **Figure 16**. It can be seen that, the dispersion of CNT/CB particles/clusters was homogeneous in the ENR matrix, whereas poor CNT/CB dispersion with strong filler-filler agglomeration was exhibited in the IR matrix as expected.

Moreover, NR-CNT/CB composite (CNT/CB 0.5/9 phr) was developed for sensor [6], it was embedded in gloves to understand its efficiency and to get a visual idea about the function of the sensors as shown in **Figure 17**.

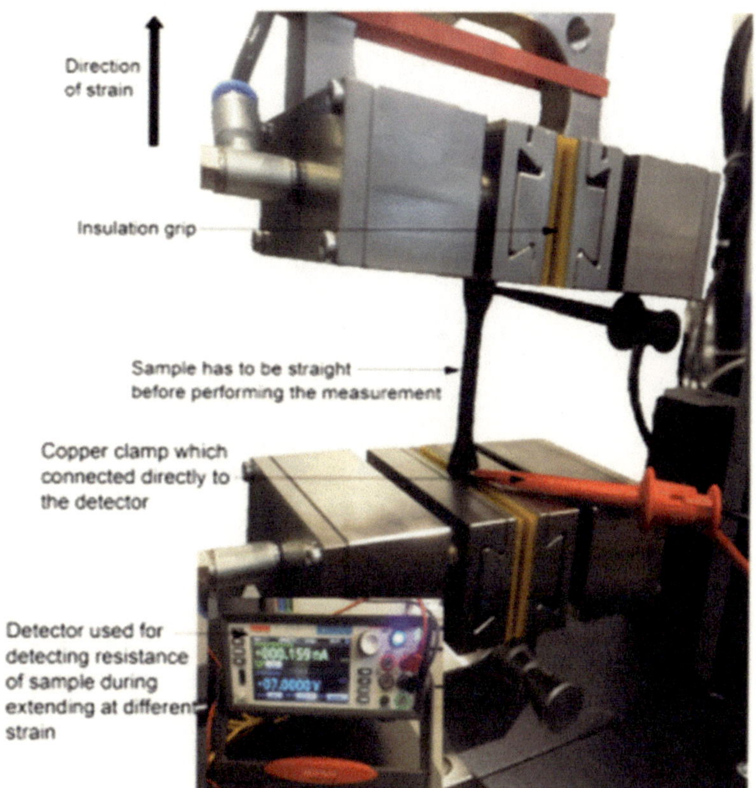

Figure 15.
Instrumental setup for measuring electrical conductivity and resistivity during mechanical tensile strain [6].

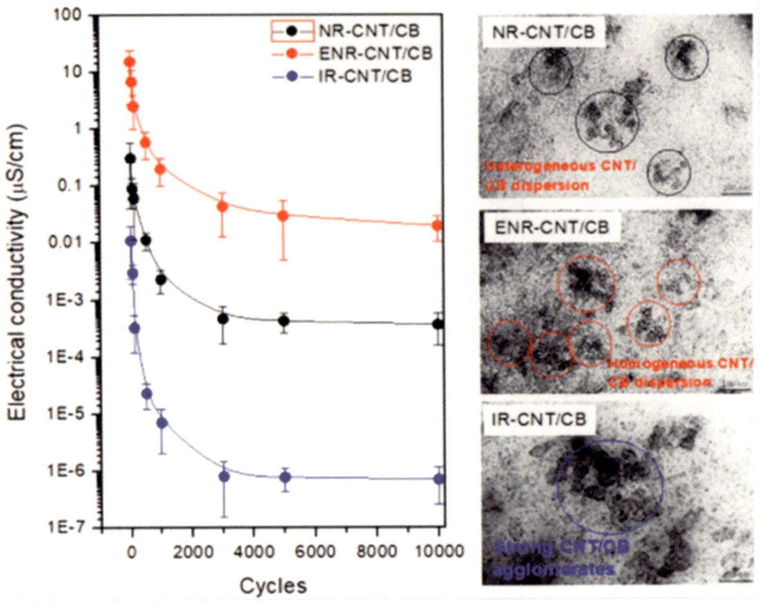

Figure 16.
Electrical conductivity of NR, ENR and IR composites with 5 phr of CNT/CB hybrid filler compared after 0, 50, 100, 300, 500, 1000, 3000, 5000 and 10000 cycles of extensional strain, together with TEM images at the same magnification of 50 kx [6].

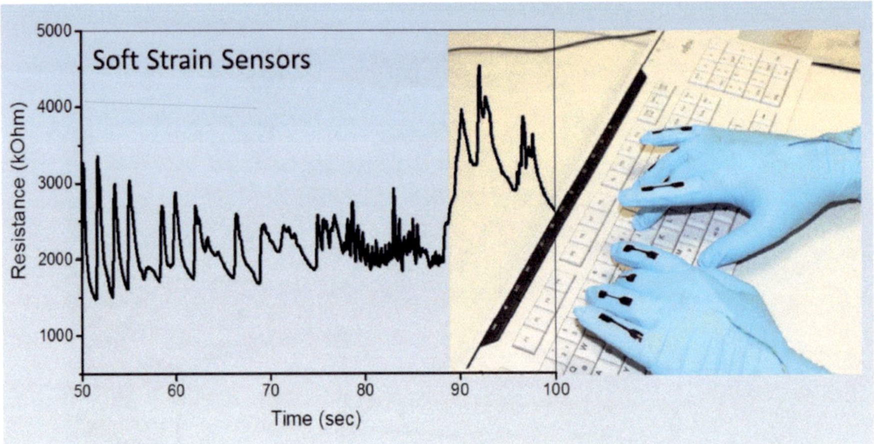

Figure 17.
Detection of finger motion and type of the motion by embodiment of conducting elastomer composite (CNT/CB 0.5/9 phr) on latex gloves. Typing the complex stretching and bending motion of sample is directly reflected in the resistance plot [6].

5. Conclusion

Carbon nanotubes (CNT) have been widely used as the reinforcing and conductive filler in NR. However, the dispersion of CNT in NR matrix is limited and always an important factor to enhance the property of NR composites. In order to obtain a conductive NR material with high quality by the formation of strong CNT networks in an insulating NR matrix is needed. The CNT networks act as electrically conducting pathways to provide electrical conductivity, but the CNT typically has a high aspect ratio and strong Van-der Waals forces that give rise to a strong agglomeration tendency. It is very difficult to form the conductive paths with in the insulating rubber matrix and this path formation between the conducting particles is a challenge to achieve proper electron tunneling.

This chapter reports several advanced methods to disperse CNTs in the NR matrix. Various aspects highlighted in this chapter include the mixing processes such as melt mixing and latex mixing methods. In addition, formations of functional groups on the surfaces of CNT using silane coupling agents (i.e., ex-situ and in-situ functionalization) as well as using a hybrid CNT are beneficial to achieve better electrical conductivity. These efforts are aimed to reduce the percolation threshold concentration in the NR composites. As mentioned in this review, latex mixing technique exhibits the formation of segregated nanotube network, which enhances the electrical conductivity of the composites. In addition, the improved interaction between CNT and NR matrix by using silane coupling agent enhances the uniformity of dispersion of CNT. It leads to reduce the percolation threshold concentration compared to the composites of NR/CNT without silane coupling agent. Moreover, the addition of secondary fillers into the composites generates new conductive hybrid filler pathways. Comparatively better conductivity is achieved by the addition of CB or AgNP or IL into the CNT polymer composites.

However, conducting composites based on electrically insulating rubber matrix have been developed for sensor applications. Change in resistivity during elongation termed as piezoresistivity can be used in sensor applications. The most commonly used rubber matrices such as NR, ENR and IR are mixed with a combination of CNT and CB fillers as a hybrid filler. The presence of linkages in the ENR composites results in the least loss of conductivity during external strain. It is found that the conductivity becomes stable after 3000 cycles. This is found to be similar to the NR-CNT/CB composite, while a few cycles are needed for IR-CNT/CB owing to the higher filler agglomeration and poor filler-rubber interactions. This is attributed to the polar chemical interactions between ENR and the functional groups on the surfaces of CNT/CB. Furthermore, the non-rubber components in NR and ENR matrices improved the filler dispersion. Finally, it can be concluded that the composite of ENR and CNT/CB are beneficial in sensor applications particularly in case of health monitoring, motion detectors, and other related products because of its cost-effectiveness and ease of processing.

Author details

Apinya Krainoi[1], Jobish Johns[2], Ekwipoo Kalkornsurapranee[3]
and Yeampon Nakaramontri[4*]

1 Department of Biotechnology, School of Bioresources and Technology, King Mongkut's University of Technology Thonburi, Bangkok, Thailand

2 Department of Physics, Rajarajeswari College of Engineering, Bangalore, India

3 Department of Materials Science and Technology, Faculty of Science, Prince of Songkla University, Hat-Yai, Thailand

4 Sustainable Polymer and Innovative Composite Materials Research Group, Department of Chemistry, King Mongkut's University of Technology Thonburi, Bangkok, Thailand

*Address all correspondence to: yeampon.nak@kmutt.ac.th

References

[1] Le H, Abhijeet S, Ilisch S, Klehm J, Henning S, Beiner M, et al. The role of linked phospholipids in the rubber-filler interaction in carbon nanotube (CNT) filled natural rubber (NR) composites. Polymer. 2014;55(18):4738-4747.

[2] Krainoi A, Kummerlöwe C, Nakaramontri Y, Vennemann N, Pichaiyut S, Wisunthorn S, et al. Influence of critical carbon nanotube loading on mechanical and electrical properties of epoxidized natural rubber nanocomposites. Polymer Testing. 2018;66:122-136.

[3] Yamada T, Hayamizu Y, Yamamoto Y, Yomogida Y, Izadi-Najafabadi A, Futaba DN, et al. A stretchable carbon nanotube strain sensor for human-motion detection. Nature nanotechnology. 2011;6(5):296.

[4] Wang Y, Wang L, Yang T, Li X, Zang X, Zhu M, et al. Wearable and highly sensitive graphene strain sensors for human motion monitoring. Advanced Functional Materials. 2014;24(29):4666-4670.

[5] Chen J, Li H, Yu Q, Hu Y, Cui X, Zhu Y, et al. Strain sensing behaviors of stretchable conductive polymer composites loaded with different dimensional conductive fillers. Composites Science and Technology. 2018;168:388-396.

[6] Natarajan TS, Eshwaran SB, Stöckelhuber KW, Wießner S, Pötschke P, Heinrich G, et al. Strong strain sensing performance of natural rubber nanocomposites. ACS applied materials & interfaces. 2017;9(5):4860-4872.

[7] Zhang XW, Pan Y, Zheng Q, Yi XS. Piezoresistance of conductor filled insulator composites. Polymer international. 2001;50(2):229-236.

[8] Zhao J, Dai K, Liu C, Zheng G, Wang B, Liu C, et al. A comparison between strain sensing behaviors of carbon black/polypropylene and carbon nanotubes/polypropylene electrically conductive composites. Composites Part A: Applied Science and Manufacturing. 2013;48:129-136.

[9] Nakaramontri Y, Nakason C, Kummerloewe C, Vennemann N. Effects of in-situ functionalization of carbon nanotubes with bis (triethoxysilylpropyl) tetrasulfide (TESPT) and 3-amino-propyltriethoxysilane (APTES) on properties of epoxidized natural rubber–carbon nanotube composites. Polymer Engineering & Science. 2015;55(11):2500-2510.

[10] GeorgeN, ChandraJ, MathiazhaganA, Joseph R. High performance natural rubber composites with conductive segregated network of multiwalled carbon nanotubes. Composites Science and Technology. 2015;116:33-40.

[11] Ma P-C, Siddiqui NA, Marom G, Kim J-K. Dispersion and functionalization of carbon nanotubes for polymer-based nanocomposites: a review. Composites Part A: Applied Science and Manufacturing. 2010;41(10):1345-1367.

[12] Sun X, Sun H, Li H, Peng H. Developing polymer composite materials: carbon nanotubes or graphene? Advanced Materials. 2013;25(37):5153-5176.

[13] Nakaramontri Y, Nakason C, Kummerlöwe C, Vennemann N. Influence of modified natural rubber on properties of natural rubber–carbon nanotube composites. Rubber Chemistry and Technology. 2015;88(2):199-218.

[14] Earp B, Simpson J, Phillips J, Grbovic D, Vidmar S, McCarthy J, et al. Electrically Conductive CNT Composites at Loadings below Theoretical Percolation Values. Nanomaterials. 2019;9(4):491.

[15] Sui G, Zhong W, Yang X, Zhao S. Processing and material characteristics of a carbon-nanotube-reinforced natural rubber. Macromolecular Materials and Engineering. 2007;292(9):1020-1026.

[16] Nah C, Lim JY, Cho BH, Hong CK, Gent AN. Reinforcing rubber with carbon nanotubes. Journal of applied polymer science. 2010;118(3):1574-1581.

[17] Atieh MA, Girun N, Ahmadun FlR, Guan C, Mahdi E, Baik D. Multi-wall carbon nanotubes/natural rubber nanocomposite. AzoNano–Online Journal of Nanotechnology. 2005;1:1-11.

[18] Bokobza L, Belin C. Effect of strain on the properties of a styrene–butadiene rubber filled with multiwall carbon nanotubes. Journal of applied polymer science. 2007;105(4):2054-2061.

[19] Bokobza L. Multiwall carbon nanotube elastomeric composites: A review. Polymer. 2007;48(17):4907-4920.

[20] Peng Z, Feng C, Luo Y, Li Y, Kong L. Self-assembled natural rubber/multi-walled carbon nanotube composites using latex compounding techniques. Carbon. 2010;48(15):4497-4503.

[21] Subramaniam K, Das A, Stöckelhuber KW, Heinrich G. Elastomer composites based on carbon nanotubes and ionic liquid. Rubber Chemistry and Technology. 2013;86(3):367-400.

[22] Nakaramontri Y, Kummerlöwe C, Nakason C, Vennemann N. The effect of surface functionalization of carbon nanotubes on properties of natural rubber/carbon nanotube composites. Polymer Composites. 2015;36(11):2113-2122.

[23] George N, Varghese GA, Joseph R. Improved mechanical and barrier properties of Natural rubber-Multiwalled carbon nanotube composites with segregated network structure. Materials Today: Proceedings. 2019;9:13-20.

[24] Krainoi A, Kummerlöwe C, Nakaramontri Y, Wisunthorn S, Vennemann N, Pichaiyut S, et al. Novel natural rubber composites based on silver nanoparticles and carbon nanotubes hybrid filler. Polymer Composites. 2020;41(2):443-458.

[25] Kaewsakul W, Sahakaro K, Dierkes WK, Noordermeer JW. Optimization of mixing conditions for silica-reinforced natural rubber tire tread compounds. Rubber chemistry and technology. 2012;85(2):277-294.

[26] Sengloyluan K, Sahakaro K, Dierkes WK, Noordermeer JW. Silica-reinforced tire tread compounds compatibilized by using epoxidized natural rubber. European polymer journal. 2014;51:69-79.

[27] Bhattacharyya S, Sinturel C, Bahloul O, Saboungi M-L, Thomas S, Salvetat J-P. Improving reinforcement of natural rubber by networking of activated carbon nanotubes. Carbon. 2008;46(7):1037-1045.

[28] Moniruzzaman M, Winey KI. Polymer nanocomposites containing carbon nanotubes. Macromolecules. 2006;39(16):5194-5205.

[29] Nakaramontri Y, Kummerlöwe C, Vennemann N, Wisunthorn S, Pichaiyut S, Nakason C. Effect of bis (triethoxysilylpropyl) tetrasulfide (TESPT) on properties of carbon nanotubes and conductive carbon black hybrid filler filled natural rubber nanocomposites. Express Polymer Letters. 2018;12(10):867-884.

[30] Nakaramontri Y, Nakason C, Kummerlöwe C, Vennemann N. Enhancement of electrical conductivity and filler dispersion of carbon nanotube filled natural rubber composites by latex mixing and in situ silanization.

Rubber Chemistry and Technology. 2016;89(2):272-291.

[31] Nakaramontri Y, Nakason C, Kummerlöwe C, Vennemann N. Enhancement of electrical conductivity and other related properties of epoxidized natural rubber/carbon nanotube composites by optimizing concentration of 3-aminopropyltriethoxy silane. Polymer Engineering & Science. 2017;57(4):381-391.

[32] Sumfleth J, Adroher XC, Schulte K. Synergistic effects in network formation and electrical properties of hybrid epoxy nanocomposites containing multi-wall carbon nanotubes and carbon black. Journal of materials science. 2009;44(12):3241-3247.

[33] Li C, Thostenson ET, Chou T-W. Dominant role of tunneling resistance in the electrical conductivity of carbon nanotube–based composites. Applied Physics Letters. 2007;91(22):223114.

[34] Ma P-C, Liu M-Y, Zhang H, Wang S-Q, Wang R, Wang K, et al. Enhanced electrical conductivity of nanocomposites containing hybrid fillers of carbon nanotubes and carbon black. ACS applied materials & interfaces. 2009;1(5):1090-1096.

[35] Zhang S, Lin L, Deng H, Gao X, Bilotti E, Peijs T, et al. Synergistic effect in conductive networks constructed with carbon nanofillers in different dimensions. Express Polym Lett. 2012;6(2):159-168.

[36] Lee J-H, Kim SK, Kim NH. Effects of the addition of multi-walled carbon nanotubes on the positive temperature coefficient characteristics of carbon-black-filled high-density polyethylene nanocomposites. Scripta Materialia. 2006;55(12):1119-1122.

[37] Dang Z-M, Shehzad K, Zha J-W, Mujahid A, Hussain T, Nie J, et al. Complementary percolation characteristics of carbon fillers based electrically percolative thermoplastic elastomer composites. Composites science and technology. 2011;72(1):28-35.

[38] Zhang X, Liang G, Chang J, Gu A, Yuan L, Zhang W. The origin of the electric and dielectric behavior of expanded graphite–carbon nanotube/ cyanate ester composites with very high dielectric constant and low dielectric loss. Carbon. 2012;50(14):4995-5007.

[39] Sumfleth J, de Almeida Prado LA, Sriyai M, Schulte K. Titania-doped multi-walled carbon nanotubes epoxy composites: Enhanced dispersion and synergistic effects in multiphase nanocomposites. Polymer. 2008;49(23):5105-5112.

[40] Ma PC, Tang BZ, Kim J-K. Effect of CNT decoration with silver nanoparticles on electrical conductivity of CNT-polymer composites. Carbon. 2008;46(11):1497-1505.

[41] Nakaramontri Y, Pichaiyut S, Wisunthorn S, Nakason C. Hybrid carbon nanotubes and conductive carbon black in natural rubber composites to enhance electrical conductivity by reducing gaps separating carbon nanotube encapsulates. European Polymer Journal. 2017;90:467-484.

[42] Krainoi A, Kummerlöwe C, Vennemann N, Nakaramontri Y, Pichaiyut S, Nakason C. Effect of carbon nanotubes decorated with silver nanoparticlesashybridfilleronproperties of natural rubber nanocomposites. Journal of Applied Polymer Science. 2019;136(13):47281.

[43] Deng F, Ito M, Noguchi T, Wang L, Ueki H, Niihara K-i, et al. Elucidation of the reinforcing mechanism in carbon nanotube/rubber nanocomposites. ACS nano. 2011;5(5):3858-3866.

[44] Fukushima T, Kosaka A, Yamamoto Y, Aimiya T, Notazawa S,

Takigawa T, et al. Dramatic effect of dispersed carbon nanotubes on the mechanical and electroconductive properties of polymers derived from ionic liquids. Small. 2006;2(4):554-560.

[45] Das A, Stöckelhuber KW, Jurk R, Fritzsche J, Klüppel M, Heinrich G. Coupling activity of ionic liquids between diene elastomers and multi-walled carbon nanotubes. Carbon. 2009;47(14):3313-3321.

[46] Subramaniam K, Das A, Simon F, Heinrich G. Networking of ionic liquid modified CNTs in SSBR. European Polymer Journal. 2013;49(2):345-352.

[47] Subramaniam K, Das A, Steinhauser D, Klüppel M, Heinrich G. Effect of ionic liquid on dielectric, mechanical and dynamic mechanical properties of multi-walled carbon nanotubes/polychloroprene rubber composites. European polymer journal. 2011;47(12):2234-2243.

[48] Marwanta E, Mizumo T, Nakamura N, Ohno H. Improved ionic conductivity of nitrile rubber/ ionic liquid composites. Polymer. 2005;46(11):3795-3800.

[49] Marzec A, Laskowska A, Boiteux G, Zaborski M, Gain O, Serghei A. The impact of imidazolium ionic liquids on the properties of nitrile rubber composites. European Polymer Journal. 2014;53:139-146.

[50] Krainoi A, Nakaramontri Y, Wisunthorn S, Pichaiyut S, Nakason C, Kummerlöwe C, et al. Influence of carbon nanotube and ionic liquid on properties of natural rubber nanocomposites. Express Polymer Letters. 2019;13(4).

[51] Shui X, Chung D. A piezoresistive carbon filament polymer-matrix composite strain sensor. Smart materials and structures. 1996;5(2):243.

[52] Kim K-S, Rhee K-Y, Lee K-H, Byun J-H, Park S-J. Rheological behaviors and mechanical properties of graphite nanoplate/carbon nanotube-filled epoxy nanocomposites. Journal of industrial and engineering chemistry. 2010;16(4):572-576.

[53] Wang J, Zhang K, Cheng Z, Lavorgna M, Xia H. Graphene/carbon black/natural rubber composites prepared by a wet compounding and latex mixing process. Plastics, Rubber and Composites. 2018;47(9):398-412.

[54] Melnykowycz M, Koll B, Scharf D, Clemens F. Comparison of piezoresistive monofilament polymer sensors. Sensors. 2014;14(1):1278-1294.

Chapter 5

Carbon Nanotubes as Reinforcing Nanomaterials for Rubbers Used in Electronics

Jabulani I. Gumede, James Carson and Shanganyane P. Hlangothi

Abstract

The field of electronics involves complex systems where the active and passive electronic devices are integrated on the rubber substrate, e.g., silicone (Q), which provides, through potting, a strong assembly of these devices on the circuit board. Several other rubbers are employed in the field to strengthen, insulate and seal the components of the electronic machines and instruments, and therefore protect them against damage. These rubbers are typically strengthened and toughened using carbon black (CB). However, due to its noticeable drawbacks, recent research in the field of rubber and electronics has suggested the use of carbon nanotubes (CNTs) as alternative reinforcing fillers to produce electronics rubber composites that do not only have enhanced electrical conductiv¬ity, thermal stability, electromagnetic interference (EMI) shielding, weatherability and insulation properties, but also offer outstanding stretchability, bendability and tear strength under frequent elastic deformation. These performances are similar for both single-walled carbon nanotubes (SWCNTs) and multi-walled carbon nanotubes (MWCNTs) in both the functional and structural composites. Although SWCNTs can result in relatively better homogeneity than MWCNTs, most rubbers often constitute MWCNTs because they are relatively cheaper. The great potential of rubber-CNTs composites being extensively used in the field of electronics is explored in this chapter.

Keywords: carbon nanotubes, electronics, rubber-carbon nanotubes composites, nanoscale filler, rubber properties

1. Introduction

Various electronic machines and instruments are available worldwide, and some of these are shown in **Table 1**. The purpose of these machines and instruments is to make different aspects of human lives easier. Components such as electronic devices (i.e., integrated circuits), wires and cables are central to their make-up [1, 4]. Some of these components are either made of rubber materials or require the use of various types of rubber for their respective functions. These include rubber sheets, grommets, tubes and seals, keypads, wire and cable rubber hoses and insulators, adhesive sealants, flat washers, boots and bellows, bumpers and tips covers, sleeves, and anti-vibration rubber mounts [4–9]. Their main function is to keep the machines and instruments

Electronic machines and instruments	Sector
Stethoscope, respiration monitors, defibrillator, glucose meter, and pacemaker.	Medical
Entertainment and navigation systems, engine, transmission, and devices for safety and driver assistance.	Automotive (automobiles)
Office gadgets, home appliances, audio and videos systems, router, automated teller machine, and barcode scanners.	Consumer
Automation and motion control robotics, power converting technological instruments, hydraulics, and photo voltaic systems.	Industrial
Data logger, hygrometer, anemometer, drifter buoy, barometer, and tipping bucket rain gauge.	Meteorological and Oceanographic
Aircraft and missile launching systems, boom barrier and radars for military, cockpit controllers, and rocket launchers for space.	Defense and Aerospace
High voltage DC transmission, excitation systems, VAR compensation, static circuit breakers, fans and boiler feed pumps, and supplementary energy systems.	Utility systems

Table 1.
Electronic machines and instruments, and sectors where they are used [1–3].

performing at their optimum best by dissipating heat, as well as insulating and sealing the electronic components; thus protecting them against shock, electromagnetic interference, very high and low temperatures, gas permeation, and exposure to dust and fluid (e.g., water, chemicals, solvents, steam, moisture and oil) intrusion that may lead to damage of the machines and instruments [1, 2, 6, 7, 9–11]. Electronic machines and instruments normally operate at different types of environmental conditions, from moderate to high-stress environment. Therefore, in addition to being thermal, gas and fluid resistant, the electronics rubber materials are required to be extraordinarily durable, stretchable and resilient, and yet be easy to use [5, 7, 9, 12].

The common rubbers that are used to manufacture the electronics rubber materials include natural rubber (NR), Ethylene Propylene Diene Monomer (EPDM), silicone (Q), styrene butadiene rubber (SBR), nitrile butadiene rubber (NBR), Fluoroelastomer (FKM/FPM), isoprene rubber (IR) and neoprene [5–7, 9, 13–17]. Generally, these rubbers have excellent elasticity and deformability, but in addition to the fact that some of them are not crystallizable under high strain, their strength and modulus, especially, could not satisfy the requirements of some electronic machines and instruments, especially those that operate at frequent vibrations and high pressures [5, 9, 13, 18–20]. Therefore, it is deemed necessary to further strengthen these rubbers, typically by adding reinforcing filler into them to yield sufficiently high mechanical properties with low hysteresis loss (heat-build up). Other common properties that are improved by reinforcement include electrical, chemical, swelling and thermal properties [9, 13, 15, 16, 21, 22].

Carbon black (CB) and silica are the most conventional fillers used for the reinforcement of rubber. However, there are many emerging fillers such as carbon nanotubes (CNTs) which offer superior reinforcing effect, properties and performances at relatively lower quantities [2, 7, 23–25]. Although there is paucity in CNTs reinforced rubbers as compared to CB reinforced rubbers, several researchers [5, 8, 9, 12, 13, 15, 16, 21, 26–31] postulated that CNTs may be considered by electronic industries in the near future for the production of CNTs-based rubber composites for strengthening, insulating and sealing the components of the electronic machines and instruments. This is due to the resultant extraordinary properties and associated performances that are typically offered by such composites, which, by definition, refers to multiphase materials that comprises of the individual or hybrid reinforcing filler and the rubber matrix or matrices [21, 24, 28, 32].

Since the discovery of CNTs, there have been several research studies aiming at understanding their structure and properties, as well as developing novel applications for them [2, 13, 21, 33]. The main attracting nature of CNTs to produce the CNTs-based rubber composites include excellent electrical conductivity, thermal stability and chemical stability, as well as the superior mechanical properties for load-bearing reinforcements in rubber composites and for structural applications [2, 22]. Rubber-CNTs composites, with strengthened stress transfer from rubber to CNTs due to their uniform dispersion in the rubber matrix and the strong rubber-CNTs interactions, are relatively lighter and flexible for easy use in the electronic machines and instruments [2, 5, 8, 9, 13, 15]. Their excellent thermal and chemical stability makes them versatile materials for improving the flame retardancy of the electronic rubber materials and for the protection of the electronic components against damage [1, 2, 15, 28, 34]. The extraordinary durability and resilience of rubber-CNTs composites makes them suitable especially for resisting wear, high pressure and vibrations; and their outstanding mechanical properties allows them to exceptionally resist abrasion, tear, high compression set and flex fatigue life that are normally due to prolonged vibrations, high pressure and compressive loads [1, 2, 28]. These properties make them more suited and applicable for use in electronics, as well as the fact that rubber-CNTs compounds are potentially cost-effective than rubber-CB compounds because of the performance of CNTs that is dominant even at smaller loaded quantities [7, 15, 35, 36]. Hence, this chapter investigated the studies that show that CNTs have great potential as the alternative reinforcing materials for rubbers used in electronics, and these rubbers are referred to as electronics rubbers in the chapter.

2. Structure and properties of carbon nanotubes

Since the discovery of carbon nanotubes (CNTs) in 1991 by Iijima using an electric arc-discharge method and transmission electron microscope (TEM), their unique atomic structure and superior properties have attracted the researchers' attention both in academia and industry [1, 22, 24, 37]. As can be seen in **Figure 1**, CNTs are typically one-dimensional quantum nanomaterials with carbon electrons in (ideally) sp^2 hybridized orbitals, and they can also be viewed as a graphene sheet that is rolled up into a nanoscale tubular form [1, 5, 28, 38]. They are commonly synthesized using visible light vaporization, arc discharge or catalytic chemical vapor deposition method, and have been categorized as the fourth allotrope of carbon, following naturally occurring types such as diamond, graphite and fullerenes [16, 24, 37]. Depending on the employed synthesis method, CNTs can be produced as individual cylinders (single-walled carbon nanotubes, SWCNTs) or as concentric tubes (multi-walled carbon nanotubes, MWCNTs), which both have the exceptionally resilient structures due to the carbon–carbon (**–C–C–**) bond and the system controlling these atomic bonds throughout the axis of the tubes [2, 39].

CNTs, due to their size and helical arrangement of graphite rings in the walls, exhibit a wide range of interesting unique properties for various potential applications [16]. They have exceptionally small diameters (several nm) and length (μ m, mm or cm) [39, 40]. Furthermore, CNTs have incredibly high aspect ratio (~10^6) and a large surface area (~100 m^2/g to 1200 m^2/g), which often allows them to form superior interaction with the polymer matrix [22, 25, 41–43]. The strength of the sp^2 -C-C- bonds gives CNTs an extremely high tensile strength (~ 150 to 180 GPa), modulus (~ 640 GPa to 1 TPa) and elasticity, and remarkable electrical conductivity, thermal (more than 1000°C) and chemical stability [1, 5, 6, 3]. These distinctive

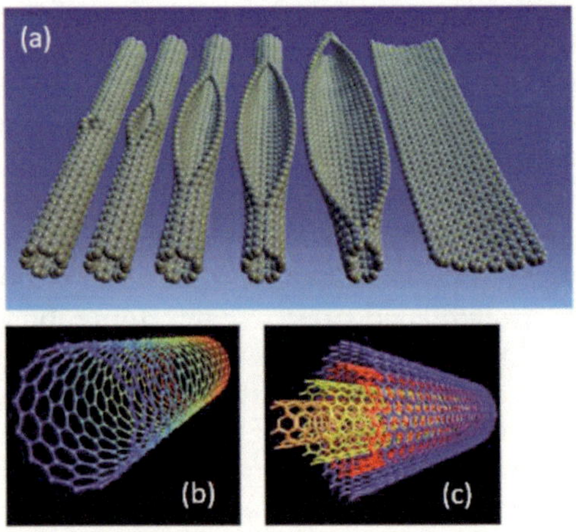

Figure 1.
Schematic illustration of CNTs: (a) carbon nano-walls, (b) SWCNTs and (c) MWCNTs (modified with permission from [28]).

properties render CNTs the ultimate nanofiller materials in nanomaterial-based rubber composites for research and industrial applications. The main attractive nature of CNTs for such composites especially in the industrial application is their fracture and deformation behavior [2]. Because of CNTs, rubber-CNTs composites can withstand high loads without showing the sign of fracture, and they can do this by switching reversibly into different morphology patterns (flattened, twisted, and buckled) on strain deformation [2].

3. Carbon nanotubes and carbon black as reinforcing fillers in carbon-based rubber composites

Generally, several fillers have been used for many years as reinforcing materials to prepare the composites mostly with enhanced mechanical properties. The mechanism of the reinforcement is believed to be both chemical and physical in nature, and the surface area and structure of the filler are its primary properties [44, 45]. The greater reinforcement effect has been reported to be given by the material with relatively smaller particle size and larger surface area [45]. Carbon black (CB) is the most widely used and most effective conventional reinforcing filler in the rubber industry because it generally enhances the mechanical properties of various rubbers [8, 18, 24, 27]. However, in addition to its environmental polluting nature, the drawbacks of using CB is that it tends to cause difficult processability due to its relatively larger particle size and high bulk viscosity of most rubber compounds [18, 24]. Additionally, its high loadings (35–45 phr) in the rubber compound formulation is a requisite for its efficiency, and this normally negatively impact the compression set and hysteresis loss (heat-build up) of some vulcanized rubber products [18, 24]. Its high loadings also cause the resultant rubber products to be relatively costly [41]. Therefore, scientific and industrial fields have been focusing on partially or completely replacing CB in rubber formulations with CNTs (single-walled and multi-walled) for the production of carbon-based composites, i.e., CNTs-natural rubber and CNTs-synthetic rubber composites, with relatively excellent properties [8, 39]. The extremely small particle size, high specific surface area and aspect ratio of this new class of fillers makes them superior to CB, and makes it relatively easy for them to uniformly disperse in the rubber matrix as individual particles [8, 28, 41]. The distances between the components of CNTs and the rubber matrix are exceptionally small and therefore, the interactions at a molecular level between CNTs and the matrix provides remarkable properties compared to conventional fillers [8].

The extraordinary properties of CNTs are the main factor that prompted a great interest in the production of CNTs-based rubber composites for a wide range of applications, including electronics [2]. The commonly used methods to prepare CNTs-based rubber composites with uniformly distributed CNTs and strong rubber-CNTs interactions include solvent/solution blending, melt blending, in-situ polymerization, latex compounding and high-shear (roll mill and internal mixer) mixing [5, 6, 40, 46–48]. The use of solvent to disperse CNTs aid in achieving good defibration and necessitate the dispersal of hydrophobic CNTs in the aqueous emulsion; hence surfactants, which typically suppress re-aggregation, are also often employed in melt blending, in-situ polymerization and latex compounding [5]. The high-shearing mixing method is often used for solid rubber and is favored for the industrial production of rubber-CNTs composites, including those used to manufacture electronics rubber materials, because it minimizes both the production time and costs [5, 28].

Since the applications of CNTs-based rubber composites are different, they are normally categorized into two kinds, i.e., functional composites and structural composites [1]. CNTs function differently in these two kinds of composites. In structural composites, they allow for the formation of structural rubber-CNTs material with easy processability, ultralight weight, and high tensile strength, elastic modulus, compression strength and stiffness [1, 36, 49, 50]. For rubber-CNTs functional composites, CNTs function by developing the electrical and thermal conductivity and chemical stability of these composites. Rubber-CNTs functional composites have shown outstanding heat resistance, chemical and swelling resistance, electrical conductivity, electromagnetic absorption and interference shielding, and high energy storing capability [8, 9, 15, 51–53].

4. Parameters affecting the properties of rubber-CNTs composites

CNTs tend to form bundles during growth due to strong van der Waals interactions between individual tubes [3, 39]. Therefore, this allows them to easily form microscale aggregates or agglomerates into a rubber matrix, hence reducing the expected improvements of the properties of the resulting composites. The extent of reinforcement effect of CNTs on rubber for the formation of CNTs-based rubber composites with superior properties is highly dependent on a variety of parameters, which normally influence the overall exploitation of the performance of rubber-CNTs composites in an intended application. These include CNTs fabrication method, ratio of CNTs to the amount of rubber matrix, entanglement state of CNTs in the rubber matrix, if CNTs are functionalized or not, functionalization method, matrix type, rubber viscosity, degree of CNTs wetting with rubber, dispersity and dispersion method, interfacial bonding, CNTs structural defects and composite processing method [13, 38, 54, 55]. These parameters are the main key for ensuring the formation of the effective load/stress transfer, normally monitored by Raman Spectroscopy, from the matrix to individual nanotube, which consequently support the effective processing of the formation of rubber-CNTs composites with optimum properties [1, 13, 21, 54, 56, 57]. Of all these, filler dispersion is the most popular parameter and is normally studied by examining the morphology of the composite, and this is achieved by employing transmission electron microscope (TEM) and scanning electron microscope (SEM). For instance, SEM micrograph can be seen in **Figure 2**, where the uniform distribution of CNTs particles throughout the silicone rubber (Q) matrix is shown in pictures (a), (b) and (c).

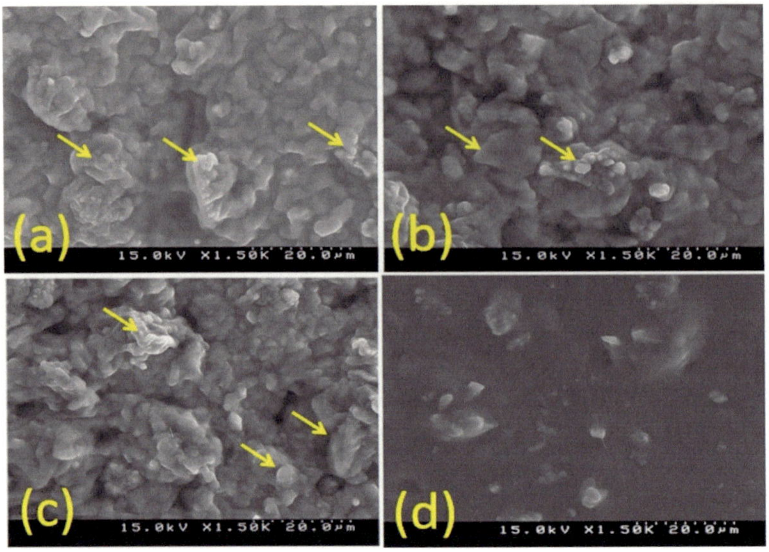

Figure 2.
SEM micrograph of Q-SWCNTs vulcanized composites: (a) Q sample of SWCNTs with 70% carbon, (b) sonicated Q sample of SWCNTs with 70% carbon, (c) Q sample of SWCNTs with 90% carbon, and (d) unfilled Q sample [58].

The parameters that define the quality of rubber-CNTs composites have also been reported to affect one another. For instance, the rubber-CNTs interaction, which generally defines the stress transfer capacity, directly affects the dispersion of CNTs in the matrix [13, 28, 59]. Also, the dispersion of CNTs in the rubber matrix and the interaction of CNTs with the matrix are highly influenced by the functionalization (surface modification) of pristine CNTs which is typically achieved by physical (non-covalent) or chemical (covalent) bonding of organic or inorganic moieties to the tubular structure of CNTs [3, 13, 28, 57, 59]. This surface modification of pristine CNTs typically result to modulation of CNTs physicochemical properties, therefore increasing their ease of dispersion and interaction, as well as processability, among different types of rubbers [38, 57]. However, even though it might possibly lead to weaker rubber-CNTs interaction, non-covalent functionalization method is more preferred than the covalent method for the production of composites as the latter tend to cause structural defects to the tubes by disturbing the π system of the graphene sheets and therefore resulting in shortened CNTs length and hence, inferior properties of the CNTs [3, 27, 60]. Optimization of the ratio of CNTs to the amount of the rubber matrix is also necessary for overcoming the inability to fully explore the properties and performance of CNTs in rubber-CNTs composites for any intended application.

5. Effect of CNTs (SWCNTs and MWCNTs) on various properties of rubber-carbon nanotubes composites for electronics

The main targets under consideration for the application of rubber-CNTs composites in electronics is to manufacture rubber materials that are extremely resistant to different temperature conditions, durable, abrasion resistant, chemical and swelling resistant, thermal resistant, high stretchable and thermo-conductive, and offer proper insulation and sealing [2, 5, 12, 13, 61]. CNTs permits the rubber-CNTs composites to maintain their mechanical strength at temperatures as high as 1200°C, therefore electronics rubber materials that comprise of CNTs would be very useful for electronics operating at high temperatures [1, 5]. Additionally, researchers [28, 62, 63] believe that rubber-CNTs composites will soon be employed as layers and coatings design to dissipate heat, and as materials that enhances flame retardancy because most rubbers are less thermally stable than CNTs. The very high chemical stability of CNTs makes the rubber-CNTs composites resistant to various solvents, oils, hydrocarbon fuels and acids or alkalis, and therefore far superior to metals as regards to corrosion resistance, and would be excellent for use in covering the electrochemical sensors, power devices and other components of various electronic machines and instruments [1, 5, 6, 28, 64]. Similarly, the high swelling stability of the rubber-CNTs composites makes them resistant to water, steam and moisture and therefore has a great potential for the manufacturing of the rubber materials that would protect the electronics materials like temperature/humidity sensors and conductive electrodes/wires [1, 6, 64]. Given the obvious possibility that water, coldness and heat, are present in the environment in which sealing and protecting rubber materials are used, rubber-CNTs composite materials would thus be best materials for electronics as they can resist hydrolysis, which may lead to degradation of rubber. This has been additionally proven by other researchers [5], where they subjected the rubber-CNTs material, with 1 wt.% CNTs content, into an environment of 280°C temperature and 6.3 MPa pressure for 3 h, and observed no change in hardness and tensile properties of the material, meaning that CNTs improve hydro-thermal resistance even at high pressure. Rubber-CNTs composites have outstanding electrical and thermal conductivity, of which the latter far exceed 400 W/Mk thermal conductivity of copper and 2200 W/Mk of diamond [5]. According to Ata [5] and Dai et al. [13], the main advantage of using rubber-CNTs composites to make stretchable electroconductive and thermo-conductive materials for electronics is that the electrical and thermal conductivity of the composite remains unaffected when the material stretches during service because CNTs form unidimensional particles in the rubber matrix, as opposed to conventional fillers. In comparison to rubber-CNTs composites, a reduction of the conductivity of the composites of conventional fillers, which are typically composites of zero-dimensional (0D) particles, with stretching, is due to the loss of contact between filler particles, as can be seen in **Figure 3** [2, 5, 50]. These conductivity properties have therefore led to a suggestion that rubber-CNTs composites can be used as wiring materials for stretchable and wearable electronic devices and instruments [5, 12, 36, 65, 66]. This has been supported by other authors [61], where they have made stretchable CNTs-based FKM conducting materials that have the potential for application in electronics such as stretchable sensors, stretchable light-emitting diodes (LEDs), and human motion monitoring.

Electronic components, like electronic devices (e.g., electronic circuits), wires and cables, in the electronic machines and instruments are subjected to damage during service. Hence, for rubber materials that typically seal, insulate and therefore protect these components, the improvement of mechanical properties, especially tensile, modulus, durability, flexibility and resilience, is a necessity. Felhös et al. [67] have used varying amounts of CNTs on hydrogenated nitrile butadiene rubber (HNBR), and measured the mechanical properties, including sliding and rolling friction, observing improved properties as the filler content was increased. They also observed that CNTs performance was better than that of silica in drying sliding and rolling. As far as the various required properties for rubber materials used in electronics are concerned, several other studies, on the reinforcement of electronics rubbers by CNTs, are shown in **Table 2**, which also shows the common applications of the stated CNTs-based rubber composites. While most of these studies are more based on enhancement of various properties of electronics rubbers, others [6, 17, 18, 24, 28, 32, 35, 41, 83, 72, 84] also compared the

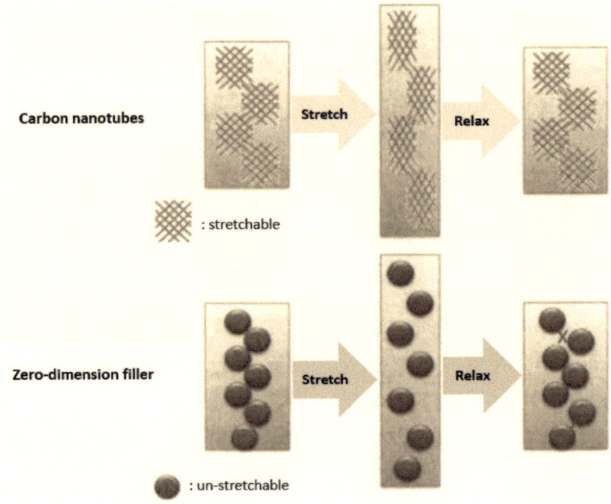

Figure 3.
Network structure in rubber composites of different fillers [5].

reinforcement effect of CNTs to that of CB, and showed that CNTs' performance is dominant even at relatively very low quantities (at or below 1 wt.%), without polluting the working environment.

It is well known that CNTs have a tubular structure of carbon atom sheets with a thickness scaled in less than few nanometers, and depending on the number of carbon atom sheets, they are often simply classified as single-walled carbon nanotubes (SWCNTs) and multi-walled carbon nanotubes (MWCNTs). The performance of SWCNTs and MWCNTs seems to be similar in both the functional and structural composites, even though it is believed that SWCNTs, in comparison with MWCNTs, are relatively more suited for the production of electronics rubber-CNTs composites since they are single-dimensional systems and therefore can easily and uniformly be dispersed in the rubber matrix and are unlikely to cause high stiffness, which might lead to high heat build-up [22, 40, 85, 86]. Nonetheless, on account of their extremely high moduli, both the SWCNTs and MWCNTs are considered fillers that provide much higher reinforcement effects than conventional fillers; hence their use in the enhancement of the matrix of most rubbers, including those that are used in the electronics [39]. It has been suggested by several researchers [6, 18, 24, 28, 32, 35, 41, 84] that the manufacturing and maintenance cost of the electronics rubber materials made of CNTs would be relatively better than those made of CB because CNTs, in addition to the fact that they are effective at very low added concentrations, are also becoming easier to fabricate and therefore cheaper to buy [5, 28, 39, 87]. As far as the cost of rubber-CNTs composites for electronics is concerned, composites made of MWCNTs would be much cheaper in comparison to those made of SWCNTs since the fabrication cost of the former is relatively cheaper than that of the latter. Nonetheless, costs are not the only determining factor on the choice of CNTs, rather the resultant properties of rubber-CNTs composites are also imperative. Therefore, it is important to understand the influence of the different types of CNTs on the properties of different types of rubbers.

5.1 Natural rubber-CNTs and isoprene rubber-CNTs composites

Natural rubber is known for its good performance in both the electronics gaskets and in insulating the electrical wires and cables because of its good elastic modulus,

Rubber-CNTs composites	Effect of CNTs on the rubber properties	Applications in the electronics sector	References
Natural rubber-CNTs	Increased ultimate tensile strength (UTS), modulus (M), tear strength, hardness and dynamic mechanical properties; decreased elongation at break (εb); and enhanced thermal stability, chemical stability, electrical (direct and alternating current) conductivity and abrasion resistance.	Electronics gaskets, insulating electrical wires and deep-sea cables, as well as electronics that are exposed to high vibrations and pressure such as washing machines, engine seals and belts.	[7, 19, 24, 33, 35, 46, 47, 51, 68]
Ethylene Propylene Diene Monomer rubber-CNTs	Increased UTS, stiffness and compression strength; decreased εb; enhanced flexibility, deformability, electrical and thermal conductivity; and improved strain sensitivity and electromagnetic shielding.	Wires, cables, power transformers, distribution automation devices, streetlights, engines and cameras.	[6, 69–71]
Silicone rubber-CNTs	Increased UTS, elastic M and hardness; decreased εb; improved thermal stability and conductivity; enhanced electrical conductivity, flexibility, and stretch-ability; and improved strain sensitivity and sensing linearity for pressure.	Cables, potting and encapsulation, particularly as sealing and insulating materials for energy transmission and distribution, electronic utility systems, high frequency communications, wearable electronics, health performance monitoring and automotive electronics.	[5, 15, 16, 50, 72–75]
Styrene butadiene rubber-CNTs	Increased UTS, stiffness, M, tear strength and hardness; decreased εb; improved thermal stability and abrasion resistance; and reduced thermogenesis and flammability.	Pressure sensors, capacitive sensors and solar cells.	[24, 41, 76–78]
Nitrile butadiene rubber-CNTs	Increased UTS, fracture strength, M, hardness, toughness strength, abrasion resistance and εb; and improved thermal stability and electrical conductivity.	Sealing for all kinds of appliances, machine tapes and power transmission belts.	[6, 18, 79–82]
Fluoroelastomer-CNTs	Increased tensile properties and stretch-ability; reduced swelling; enhanced strain/pressure sensitivity and electrical conductivity.	Electric wire and power cable coverings, Electronic circuits and dynamic pressure monitoring in electronics.	[6, 61]

Table 2.
Studies about the effect of CNTs on various properties of some rubbers used in electronics.

fracture energy and dielectric strength; therefore, it has been used in portable electronics and distributed sensors [88, 89]. But this is typically only possible when it is reinforced. A shift to CNTs from traditional fillers like CB has allowed the production of NR compounds with enhanced properties, especially because it generally fails at higher temperatures (above 100°C) and it has low oil and fuel resistance (swells in oils and fuels) [90]. Through the high-shear mixing method, Azam et al. [51] have prepared the NR-SWCNTs composites to study the effect of SWCNTs on the tensile, hardness and thermal properties of NR. They observed a reduction in the ultimate tensile strength (UTS) and elongation at break (εb) and an increase in tensile moduli (M), hardness and thermal property. The unusual decrease in the UTS of NR with increasing SWCNTs content was suspected to have been caused by the agglomeration of SWCNTs particles in the NR matrix and/or the physical contact between adjacent agglomerates. Additionally, the obtained increase in M was reported to be due to an increase in crosslink density, which resulted from increased swelling resistance of the resultant composites and good distribution of SWCNTs into the NR matrix [51]. The uniform distribution of SWCNTs resulted to limited chain movement of NR during deformation, and therefore resulted in high M, which consequently increased hardness of the rubber compounds [51]. The obtained increase in thermal properties was attributed to physical adsorption and good NR-SWCNTs chemical interaction [51]. Gumede et al. [47] used an internal mixer to prepare the NR-SWCNTs composites, but also passed the compounds through the two-roll mixing mill, which could be a reason why they observed an increase in UTS with an increase in SWCNTs content up to 0.1 wt.%. Unlike authors [47, 51], Anoop et al. [46] prepared the NR-SWCNTs composites through the latex compounding method, also employing the surfactant to improve the dispersion of SWCNTs in the NR matrix, and observed similar results to the authors [47, 51], as far as the tensile properties are concerned, and an increase in tear strength and electrical conductivity, while thermal properties remained unchanged after adding SWCNTs. These findings were attributed to good interfacial bonding between NR and SWCNTs [46]. Other researchers [19, 24, 33, 68] used MWCNTs instead of SWCNTs to prepare the NR-CNTs composites, and these studies have shown that MWCNTs are as effective as the SWCNTs for the enhancement of NR properties. For instance, Jose et al. [33] used a two-roll mixing mill to prepare NR-MWCNTs composites, and found that MWCNTs were uniformly dispersed into the NR matrix (*see* **Figure 4**), which resulted to enhanced thermal, mechanical and electrical conductivity properties of NR.

Isoprene rubber (IR) has applications that are similar to those of NR; e.g., it is typically used to manufacture the deep-sea cables insulating materials [90]. This is because IR has good electromagnetic interference (EMI) shielding property, which is important for rubber application in the electronics sector, and has been found to be proportional to electrical conductivity properties, i.e., the higher the electrical

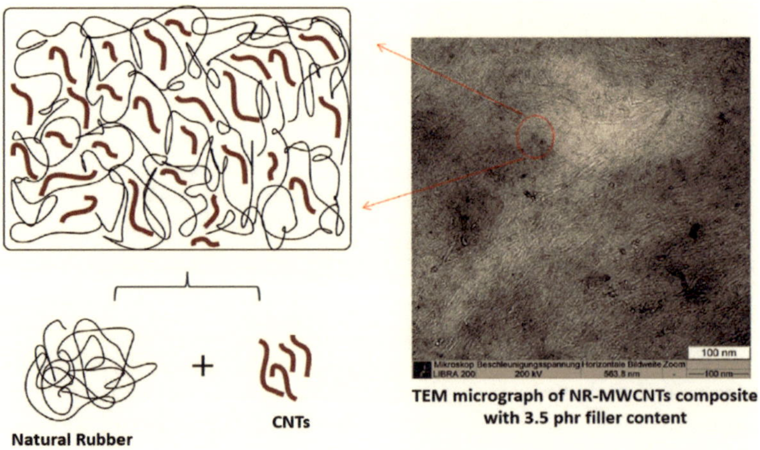

Figure 4.
Schematic representation of uniformly dispersed MWCNTs into NR matrix [33].

conductivity values, the higher the EMI shielding effectiveness [17]. IR is also employed in manufacturing the anti-vibration mounts, drive couplings and bearings; therefore, it is the best rubber especially for electronics that are exposed to high vibrations and pressure, including washing machines, engine seals and belts, as well as machines and instruments that supply power through mechanical forces [17, 90]. Wang et al. [17] have shown that the electronics rubber materials that are made of IR and CNTs exhibit high flexibility, and outstanding mechanical, electrical conductivity and EMI shielding properties, and this is mainly because CNTs are one-dimensional (1D) materials in comparison to the traditional filler (i.e., CB) that is zero-dimensional. They explained that the CNTs properties were successfully explored because of the excellent uniform dispersion of these nanomaterials into the IR matrix, and the SEM micrograph, including those of CB, can be seen in **Figure 5**. While CNTs exhibited an outstanding uniform dispersion into the IR matrix even at high content (20 wt.%), some small spherical or cluster-like aggregations for CB-IR composites were observed.

5.2 Ethylene propylene diene monomer rubber-CNTs composites

Ethylene Propylene Diene Monomer (EPDM) rubber is well-known for its outdoor applications. Considering the fact that it is the most water-resistant rubber and capable of resisting failure (abrasion, tear and degradation) during harsh weather conditions, i.e., ozone, UV ray, low-high temperature, steam and flame exposure, it is regarded as the best rubber for making seals, hoses, isolators, gaskets, roll covers, tubes, wires and cables especially for outdoor electronics, including power transformers, distribution automation devices, street lights, engines and cameras [90]. CNTs have been shown to significantly further improve the properties of EPDM for potential electronics applications. Bizhani et al. [69] prepared the EPDM-MWCNTs composite foams, as illustrated by **Figure 6**, to make an industrially scalable lightweight rubber material. Based on the cryofracture surfaces of the EPDM-MWCNTs composites that were visualized using SEM, as can be seen in **Figure 7**, to study the nature of MWCNTs dispersion in the EPDM matrix, the authors claimed to have successfully achieved EPDM-MWCNTs composite foams with a good interfacial interaction.

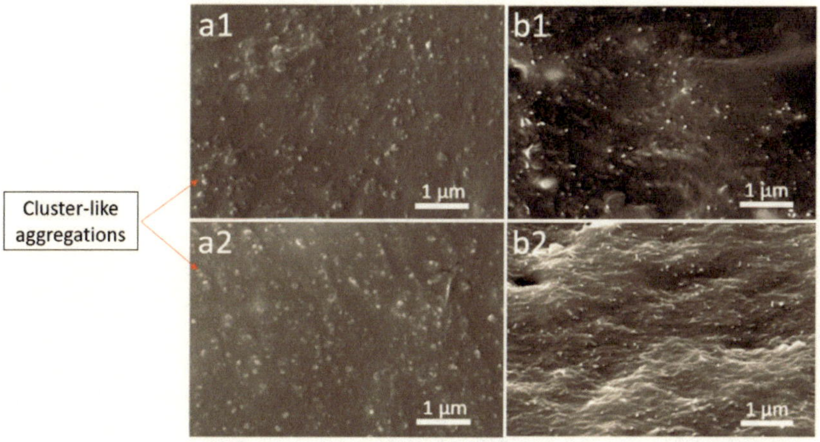

Figure 5.
SEM micrograph of IR-based composites: (a1) IR-CB composite with 10 wt.% CB content, (b1) IR-CNTs composite with 10 wt.% CNTs content, (a2) IR-CB composite with 20 wt.% CB content and (b2) IR-CNTs composite with 20 wt.% CNTs content [17].

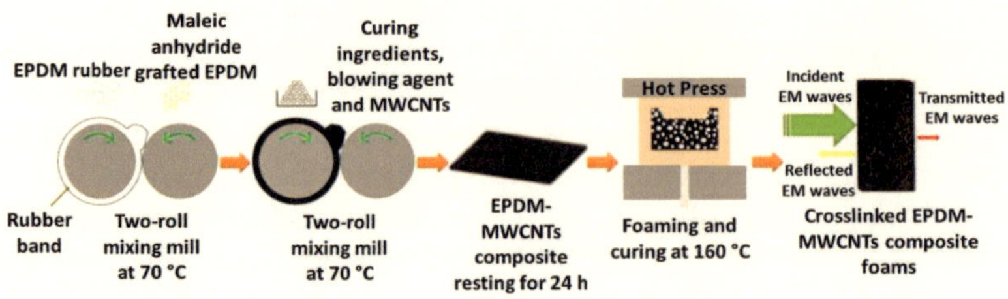

Figure 6.
Preparation of vulcanized EPDM-MWCNTs composite foams [69].

Due to the development of a good three-dimensional (3D) interconnected network between MWCNTs and EPDM matrix, the authors [69] found that MWCNTs enhanced thermal conductivity (up to 0.2 W/m·K), electrical conductivity (up to 2.7 $\times 10^{-4}$ S/cm) and EMI shielding efficiency (up to 45 dB) properties of the EPDM-MWCNTs foams, and that they do not significantly deteriorate with continuous deformation through bending, as can be seen in **Figure 8**. These properties, as well as observed high flexibility and lightweight wave absorber capability, of the EPDM-MWCNTs composite foams indicate that these foams may be best suited for several applications, including lightweight portable devices like cell phones.

Researchers [6, 70, 71] have also used high-shear mixing method to prepare the EPDM-MWCNTs composites and found that MWCNTs significantly increased the UTS and M, while reducing the εb of EPDM, due to both the good uniform distribution of MWCNTs within the EPDM matrix and the effective EPDM-MWCNTs bonding. Chougule et al. [6] explained that functional groups that are randomly orientated on the MWCNTs surface typically impacts both the level and type of interfacial bonding between MWCNTs and the EPDM matrix. MWCNTs were found to cause the swelling behavior of EPDM to decrease while increasing its electrical conductivity [6, 70]. Additionally, other authors [70] obtained the strain response that showed piezoresistivity behavior under deformation, and this indicates that the EPDM-MWCNTs composite materials have a great potential for being used as flexible strain-sensitive materials. This has also been shown by Haj-Ali et al. [91].

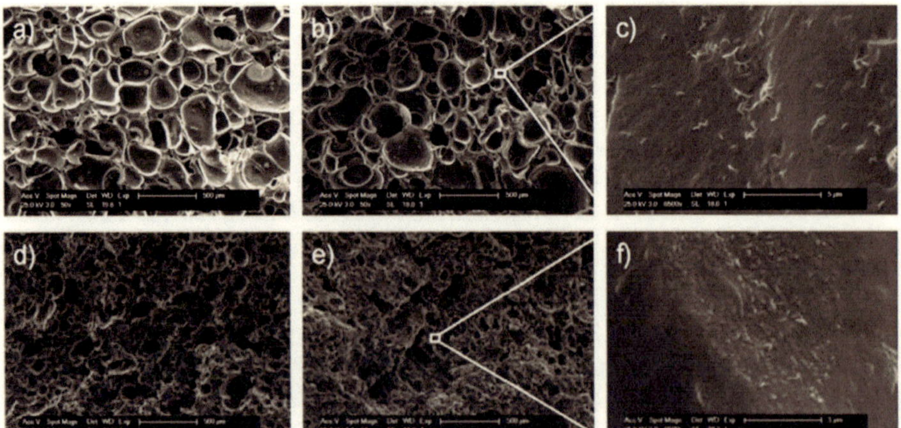

Figure 7.
SEM micrograph of EPDM-MWCNTs composite foams: (a) with 0, (b) with 2, (d) with 6 and (e) with 10 phr MWCNTs content. (c) and (f) are micrograph of filler within the morphology [69].

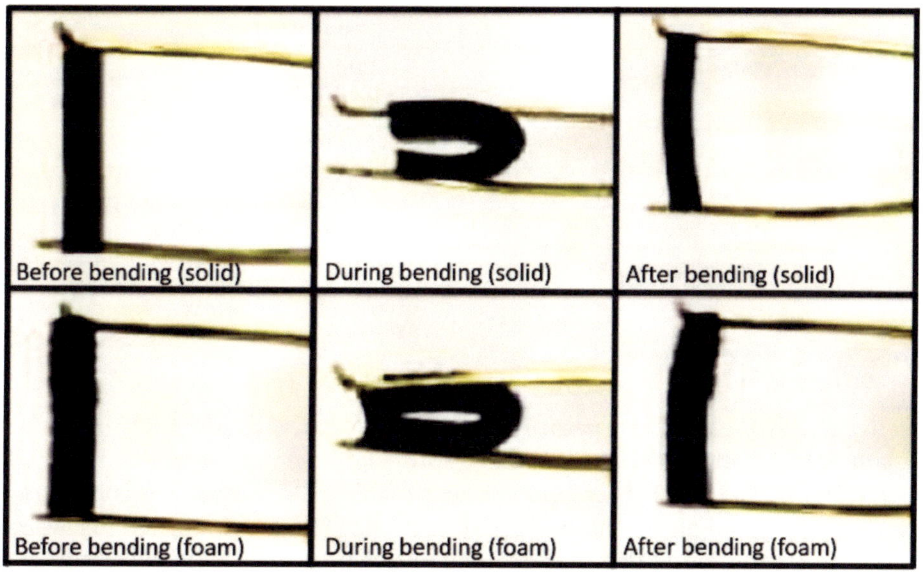

Figure 8.
Vulcanized solid and foamed EPDM-MWCNTs composites before and after bending [69].

5.3 Silicone rubber-CNTs composites

Silicone rubber (Q) is considered an inorganic–organic hybrid polymer because of the inorganic silicon-oxygen main chain with two organic groups bonded to each silicone center. As can be seen in **Table 3**, silicone rubbers can be divided based on their pendant group structure, which function to improve the natural properties of Q. Its major application is in the electronics sector mainly because of its non-toxicity, aging resistance at high temperatures, good flame resistance, chemical stability, electrical insulating and weatherability [90, 92].

CNTs are typically incorporated into Q to mainly enhance the mechanical properties, surface hydrophobicity, thermal and electrical conductivity of the resulting Q-CNTs composites. These composites, particularly the ones with liquid Q, have been reported to be very useful in high-voltage indoor and outdoor systems, and CNTs function by extending the service life of Q in such applications [92, 93]. Since Q-CNTs composites are heat, fire and chemically resistant with excellent electrical conductivity and weatherability, they can ideally be employed in the cable, potting and encapsulation sector, particularly as sealing and insulating materials for energy transmission and distribution, electronic utility systems, high frequency communications, wearable electronics, soft robotics, health and sport performance monitoring, oil drilling and automotive electronics [74, 75, 92–95]. Additionally, silicon-CNTs micropatterns can be fabricated and used in biomedical and chemical sensors, tissue engineering, drug screening, and optical devices [92]. Yanagizawa et al. [96] have prepared Q-CNTs composites and studied the effect of these nanomaterials on their water repellency. The authors observed an increase in contact angle on incorporation of just 1 wt.% of CNTs content, and that CNTs significantly improved repellency. This study aimed mainly at making the CNTs-based rubber roofing materials that are resistant to snow build-up in regions of high snowfall [5]. Li et al. [15] and Bannych et al. [75] observed a rapid increase in electrical conductivity of the Q-MWCNTs composites, after which it slowly changed when the MWCNTs content was above 1 wt.%. They reported that these trends are due to MWCNTs having a large aspect ratio, and therefore a small quantity in the rubber matrix can form an

Type	Pendant group	Chemical formula
MQ	Methyl	CH_3
PMQ	Phenyl	C_6H_5
VMQ	Vinyl	$CH_2{=}CH$
PVMQ	vinyl, phenyl	$CH_2{=}CH$, C_6H_5
FMQ	Trifluoropropyl	$CF_3CH_2CH_2$
FMVQ	Vinyl, trifluoropropyl	$CH_2{=}CH$, $CF_3CH_2CH_2$

Table 3.
Different types of silicone rubbers [90].

effective effectual conductive path. Typically, the electrical conductivity of the insulator is less than 10^{-8} S/m while that of the conductor is approximately 10^{5} S/m [15, 97]. Hence, Q-MWCNTs composites are suitable to be used as semiconductor materials [15]. Based on this suggestion, the authors [15] also conducted the Seebeck coefficient test and found that the Seebeck coefficient of the Q-MWCNTs slightly decreases with an increase in MWCNTs content, and it decreases with an increase in electrical conductivity (*see* **Figure 9**). The thermal conductivity was found to increase with an increase in MWCNTs content. Although some aggregates of MWCNTs are seen within the Q matrix in the SEM micrograph of the prepared Q-MWCNTs composite (*see* **Figure 10**), an increase in electrical and thermal conductivity has been attributed to uniform distribution of MWCNTs in Q matrix and the good interfacial bonding between MWCNTs and Q matrix [15]. Bright filament-like substances that appear on this SEM micrograph are due to MWCNTs.

Kim et al. [50] successfully prepared, *via* melt mixing method, a highly stretchable and conductive Q-SWCNTs gel composite as illustrated in **Figure 11**. The authors firstly produced the SWCNTs gels by grounding, using a mortar and pestle,

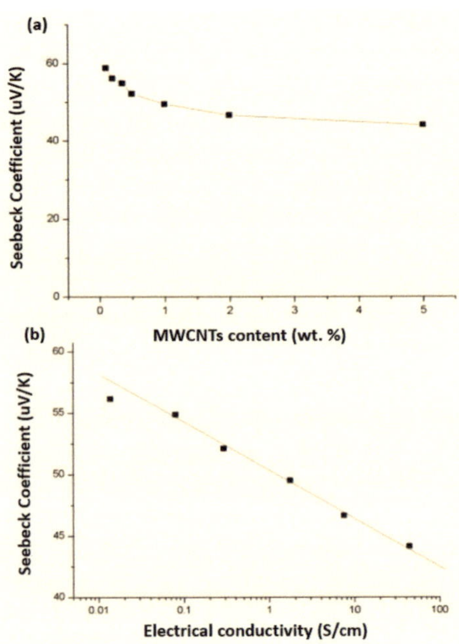

Figure 9.
Seebeck coefficient results: (a) dependent on the MWCNTs content and (b) dependent on electrical conductivity [15].

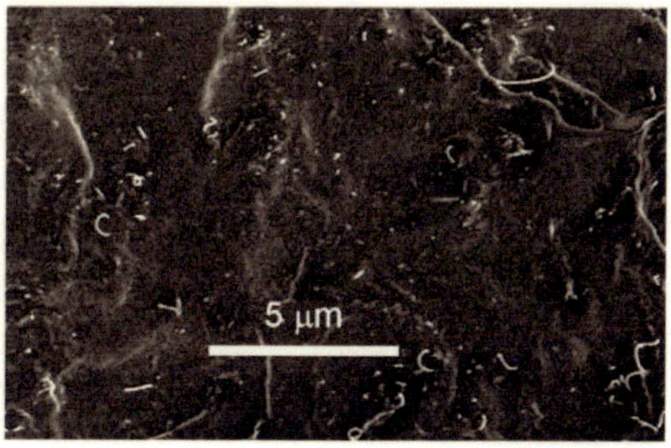

Figure 10.
SEM micrograph of Q-MWCNTs composite with 1 wt.% MWCNTs content [15].

the SWCNTs with a room temperature ionic liquid (IL) called 1-butyl-3-methylimidazolium bis(trifluoromethylsulfonyl) imide. Using an ultrasonic bath, gels were immersed in a solvent (toluene) for about 1 h to produce a solution of about 1 mg/ml concentration, after which it was homogenized with Q by stirring for about 3 h. The electrodes were made by spraying the Q-SWCNTs-IL solution onto acrylic elastomeric substrates through a contact mask and drying the samples on a hot plate for 3 h and in an ambient temperature vacuum for 24 h. The Q-SWCNTs composite electrodes were also treated with nitric acid by placing them for 30 min in a saturated acid vapor environment of 70°C, after which they were subjected to a vacuum oven at 25°C for 24 h.

The uniform dispersion of conducting SWCNTs within the Q matrix was shown by the microstructural analyses presented by **Figure 12**. This led to obtaining a composite material that is stretchable for three times its length, while maintaining its high electrical conductivity (18 S/cm) even after prolonged and continuous deformation. Hence, such materials can be best suited for wearable and stretchable conductors and strain sensors that need a constant conductivity when an intense deformation is applied [36, 50, 98].

Katihabwa et al. [73] and Shang et al. [72] also studied the effect of CNTs, particularly the multi-walled type, on the mechanical, thermal and electrical properties of the Q. Although different composite preparation methods were employed, the authors achieved a uniform dispersion of MWCNTs into the Q matrix which consequently led to the enhancement of the studied properties with an increase in the MWCNTs content in the composites. Shang et al. [72] also studied the relative resistance change (R/R_0) of the Q-MWCNTs composites when applied to elastic

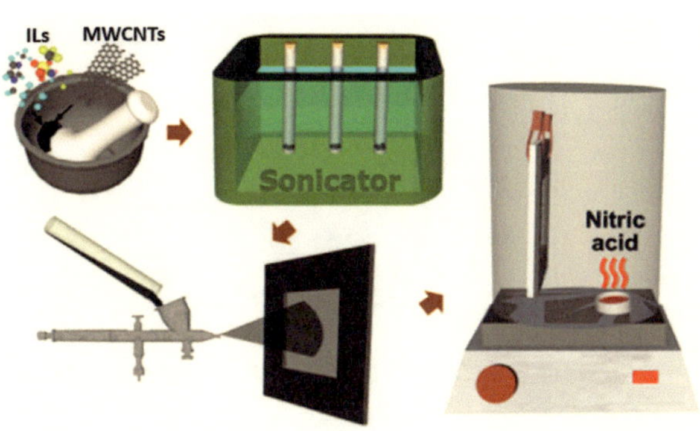

Figure 11.
Schematic representation of the preparation of Q-SWCNTs composite electrodes [50].

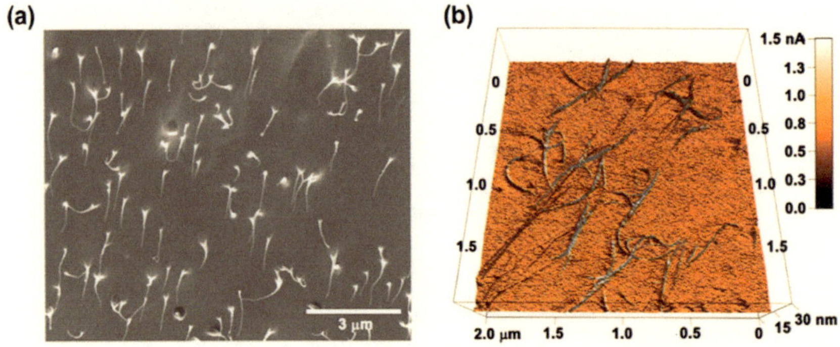

Figure 12.
Microstructural analyses of fractured regions of Q-SWCNTs composite sample with 4 wt.% SWCNTs content: (a) SEM micrograph and (b) atomic force microscope (AFM) micrograph (Amplitude represented in height difference, and current represented in color gradient) [50].

deformation (bending and twisting) of different angles. It can be seen in **Figure 13** that as the MWCNTs content reached 8 wt.% (and higher), the R/R_0 change values were smaller, meaning that there was a firm and continuous MWCNTs conducting network within the rubber matrix [72]. Hence, the prepared Q-MWCNTs composites have a great potential in the field of conductive elastomer or pressure sensors [72]. The uniform distribution of MWCNTs within Q matrix was achieved regardless of the filler content, as can be seen in **Figure 14(a)-(d)**, and this was reported to be due to the employed compatibilizer, i.e., chitosan salt, that increased the interactions between MWCNTs and Q matrix [72].

5.4 Styrene butadiene rubber-CNTs composites

Styrene butadiene rubber (SBR) has generally been employed in electronics sector because of its good elasticity and resistance to radiation, abrasion, aging, weather and ozone. However, its properties are strongly reliant on the reinforcements [90].

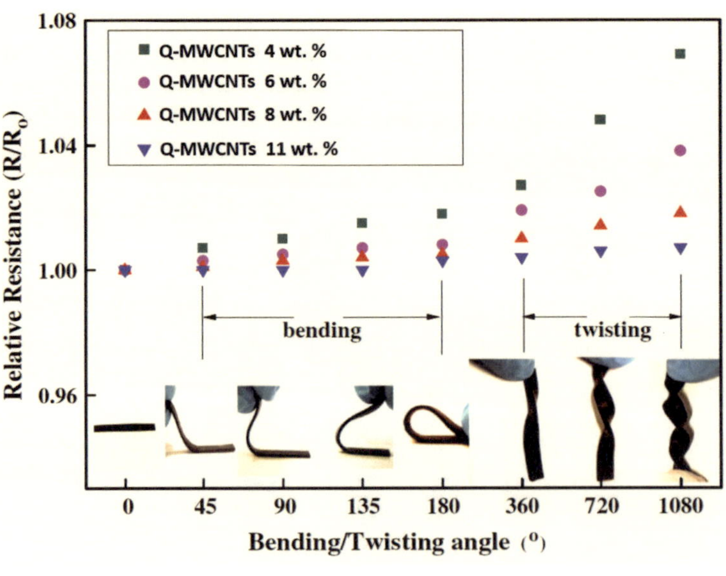

Figure 13.
Relative resistance change of the Q-MWCNTs composites at different deformation angles [72].

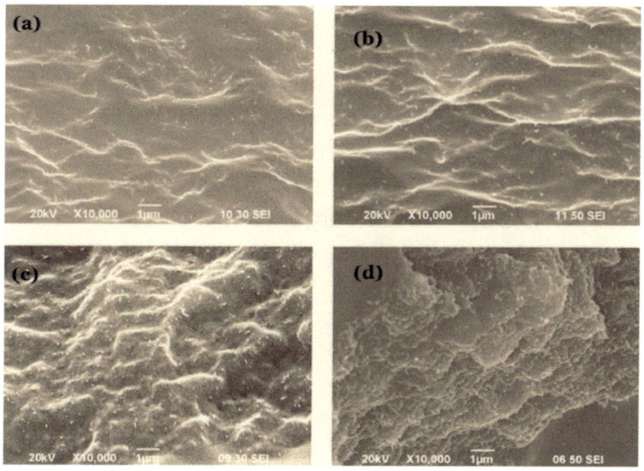

Figure 14.
SEM micrograph of Q-MWCNTs composites: (a) with 4, (b) with 6, (c) with 8 and (d) with 11 wt.% MWCNTs content [72].

Researchers [24, 41, 76, 77, 83] have reinforced the NR-SBR blend, and the results showed that MWCNTs increased the UTS, M, storage modulus, and thermal stability of the blend. This could be due to good dispersion of MWCNTs into the blend and a strong interaction between MWCNTs and the matrices of both rubbers, indicating that MWCNTs have good affinity for both NR and SBR. Gao et al. [24] also reported an improvement of the abrasion resistance of the NR-SBR (80–20 phr) blend with 5 phr MWCNTs content, which was due to the synergistic effect of MWCNTs. The authors [24] showed, by SEM and TEM (*see* **Figure 15(a)** and **(b)**), that MWCNTs can come into contact with each other and that they were uniformly distributed within NR-SBR blend matrix; meaning that a good interface cohesion between MWCNTs and the blend matrix was successfully achieved, which consequently led to obtained properties. Additionally, it was reported that, based on TEM micrograph in **Figure 15(b)**, MWCNTs can form the bridges between the CB aggregations if the CNTs/CB hybrid filler is used for rubber reinforcement [24].

Rather than blending rubbers, Liu et al. [78] mixed a reduced graphene oxide (rGO) with MWCNTs to form rGO-MWCNTs hybrid filler, which was used to prepare SBR-rGO-MWCNTs composites, as illustrated in **Figure 16**. According to the authors [78], the employed hydrothermal step facilitated the prevention of restacking of graphene sheets and agglomeration of MWCNTs, and therefore

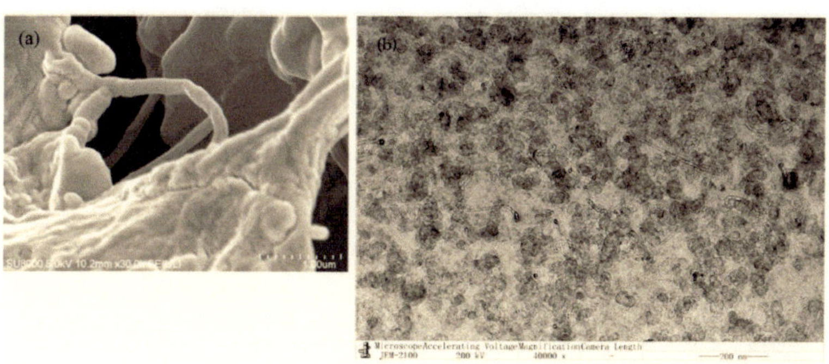

Figure 15.
Electron microscope micrographs: (a) SEM image of NR-SBR-MWCNTs composite with 5 phr filler content and (b) TEM image of NR-SBR-MWCNTs-CB composite with 5 phr MWCNTs and 27.5 phr CB content [24].

obtaining a significant increase in electrical conductivity and thermal stability of the SBR-rGO-MWCNTs composites with an increase in the rGO-MWCNTs hybrid filler content. The distribution of rGO-MWCNTs hybrid filler within SBR matrix can be seen in **Figure 17**. MWCNTs content of 10.4 wt.% gave about 3.62 S/cm of electrical conductivity of the composite, of which this was 14 orders of magnitude higher than that of unreinforced SBR. Additionally, it was reported that SBR-rGO-MWCNTs composites can retain high electrical conductivity mostly under low tensile strain. Due to the one-and two-dimensional interconnected network, formed through MWCNTs bonding with rGO sheets, SBR-rGO-MWCNTs composites can be tailored for electronics that require highly conductive and stretchable rubber materials [78].

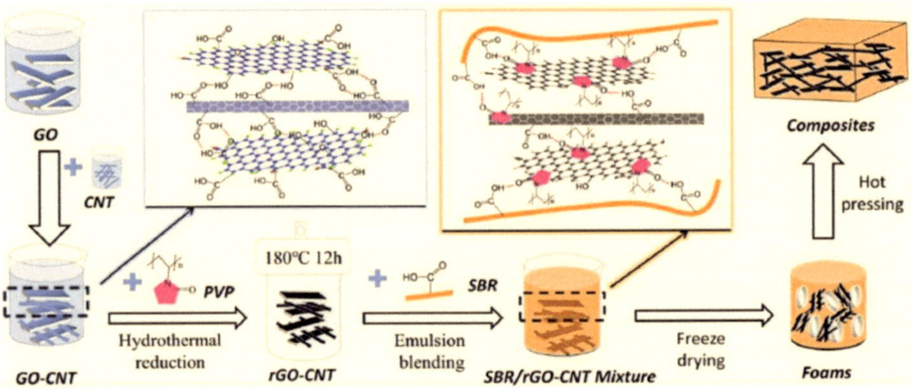

Figure 16.
Fabrication of rGO-MWCNTs hybrid and SBR-rGO-MWCNTs composites [78].

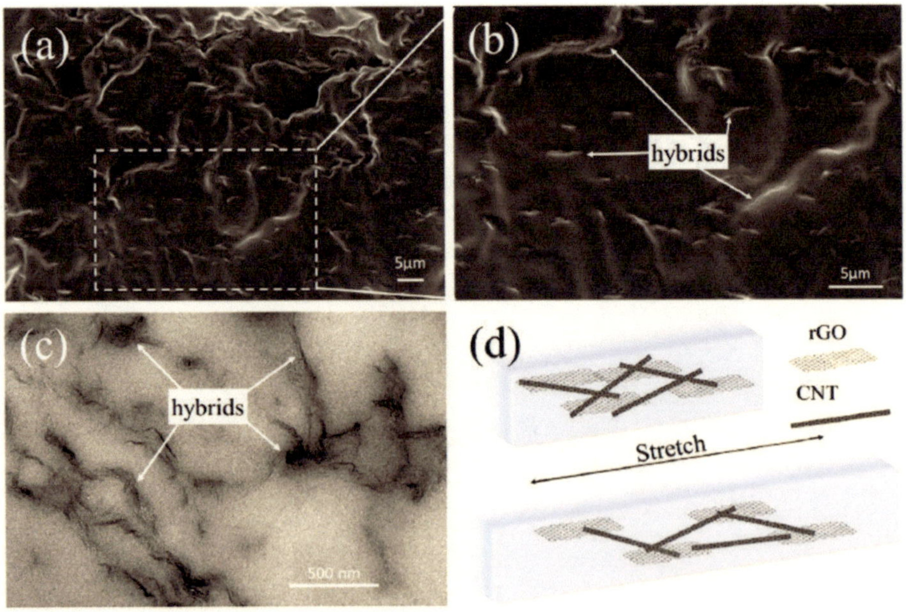

Figure 17.
(a, b) SEM and (c) TEM images of SBR-rGO-MWCNTs composite and (d) schematic representation of the redistribution of the filler hybrids within SBR composite under elastic deformation [78].

5.5 Nitrile butadiene rubber-CNTs composites

Nitrile butadiene rubber (NBR) is commonly used in making hoses, joints, sealing and roll covering materials. However, there is a growing demand for its use in electronics applications. NBR can be blended with other rubbers to improve its thermal stability and gas resistance, as well as weatherability [90]. Similar to Kim et al. [50], Wang et al. [82] also employed an ionic liquid (IL) (1-aminoethyl-3methylimidazolium bis ((trifluoromethyl) sulfonyl) imide), as well as polydopamine (PDA) and (3-Aminopropyl) triethoxysilane (KH550) to functionalize the carboxylated MWCNTs, therefore overcoming their agglomeration, and to promote the stronger carboxylated NBR-MWCNTs (XNBR-MWCNTs) interfacial bonding, which facilitated the formation of the best mechanical and damping properties (*see* **Figure 18**). The effect of functionalizing MWCNTs before they are incorporated into rubber, for the fabrication of nanomaterials-based rubber composites, were visualized by SEM micrograph, and this can be seen in **Figure 19(a)-(h)**. Agglomerated MWCNTs are clearly seen in **Figure 19(a)**, which might have indicated that pristine MWCNTs disperses poorly in XNBR matrix. Substantial crack orientations that are seen in **Figure 19(a)** are the representation of a reduction in toughness of the XNBR-pristine MWCNTs composite and explains an obtained reduction in damping (loss tan δ) properties [82]. PDA decreased the size and amount of MWCNTs agglomeration in XNBR matrix (*see* **Figure 19(c)**, and this was due to hydrogen bonding between MWCNTs-PDA and XNBR matrix [82, 99]. The authors [82] reported that MWCNTs-KH550 (**Figure 19(e)**) and MWCNTs-IL (**Figure 19(g)**) had relatively stronger interaction with XNBR, and therefore the large-scale agglomeration was prevented. XNBR-MWCNTs-KH550 (**Figure 19(e)**) and XNBR-MWCNTs-IL (**Figure 19(g)**) composites showed only small cracks in the cross-section of XNBR, which indicated that the XNBR toughness was improved by functionalizing MWCNTs before they were incorporated into rubber matrix. Storage modulus of the prepared XNBR-MWCNTs composites was seen to increase by 80% (from 1392 to 2488 MPa) with the incorporation of 2.2 wt.% CNTs-KH550, while the UTS increased from 0.32 to 0.68 MPa by 110% with 3.0 wt.% CNTs-IL.

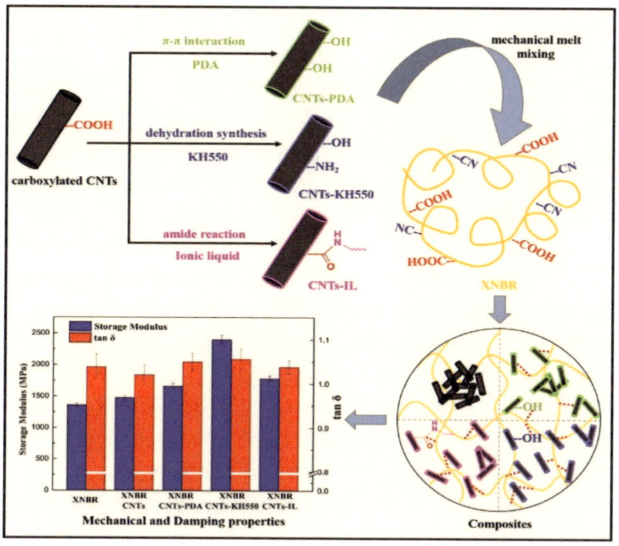

Figure 18.
Functionalization of MWCNTs and preparation of XNBR-MWCNTs composites with their studied mechanical and damping properties [82].

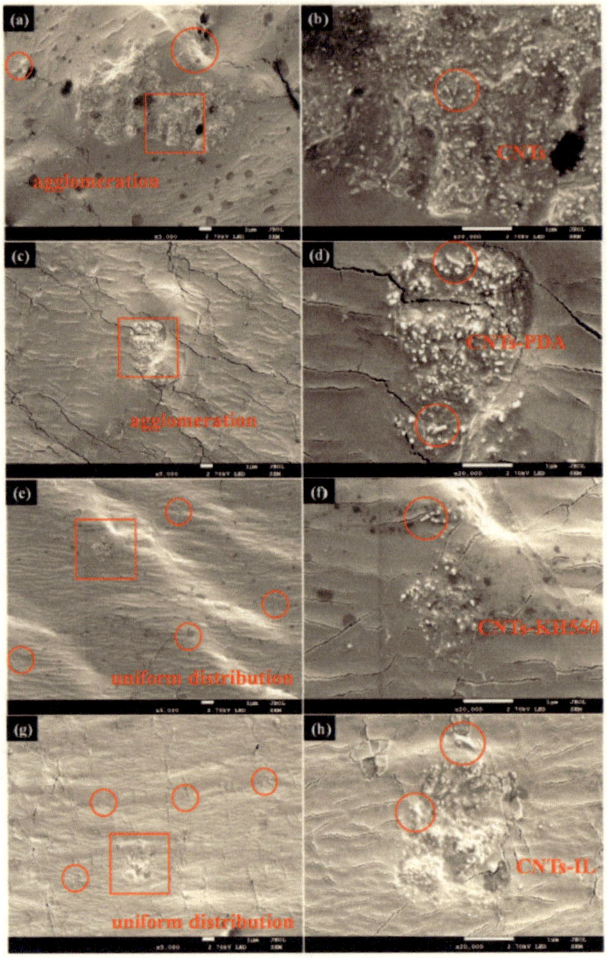

Figure 19.
Cross-sectional SEM micrograph of XNBR-MWCNTs (3 wt.%) composites with different functionalizing materials: (a, b) XNBR-MWCNTs, (c, d) XNBR-MWCNTs-PDA, (e, f) XNBR-MWCNTs-KH550 and (g, h) XNBR-MWCNTs-IL [82].

Although other researchers have reported good results of CNTs-based composites that constitute the pristine CNTs, the authors [82] have reported that the failure of damping performance that is often encountered in the field of composites, due to agglomeration of CNTs, can be suppressed by functionalizing the MWCNTs prior the fabrication of MWCNTs-based rubber composites, and this will widen the scope of the application of NBR-MWCNTs composites in the field of electronics.

Shao et al. [79] observed an improvement in UTS and hardness to about 39% and 101%, respectively, after adding CNTs into NBR, while the εb maintained at high CNTs contents over 100%. They also observed a significant reduction of volume resistivity and an increase in dielectric constant and dielectric loss with an increased CNTs content in the NBR-based nanocomposites. Additionally, the electrical percolation threshold obtained was low (1.5 pph). These results indicated the development of three-dimensional conductive networks in the composites, which meant that NBR-CNTs nanocomposites can provide insulation service to electronics [7, 79]. The presence of MWCNTs in the prepared, *via* melt and high-shear mixing methods, NBR-MWCNTs composites reduced the swelling behavior while increasing the crosslink density of the composites; consequently increasing their UTS, M, hardness, abrasion resistance and electrical conductivity (with corresponding low percolation threshold value), and significantly decreasing their volume resistivity [6, 18, 80, 81]. These results have been attributed to uniform distribution of MWCNTs into the NBR matrix and the superior interfacial bonding between MWCNTs and the host matrix (NBR) which are due to large aspect ratio of the MWCNTs.

5.6 Fluorocarbon rubber-CNTs composites

Fluorocarbon rubbers, particularly the fluoroelastomer (FKM/FPM), are popular in electronics sector mainly because of their good heat, chemical and abrasion resistance; and does a perfect job for sealing the electronics materials. Seo et al. [100] have proposed a solvent-free encapsulation method to produce the FKM-SWCNTs composite layers that can help protect the electronic components from the physisorbed moisture, water/oxygen molecules. They suggested that this method, with elastomeric poly(vinylidene fluoride-co-hexafluoropropylene) (e-PVDF-HFP) film lamination, can potentially provide the cost-effective, large-area processable and highly dependable SWCNTs-based thin-film transistors in electronics applications. To study the hydrophilicity and hydrophobicity of each layer of SWCNTs-based thin-film transistors, the contact angle measurements of deionized (DI) water on the poly-L-lysine (PLL) solution-treated layer, SWCNTs-deposited layer, and e-PVDF-HFP layer were examined as illustrated by **Figure 20(b)**. The layer that was pre-treated with PLL solution appeared to be hydrophilic with an angle of 22.3°, and this was attributed to the hydroxyl and amine groups on the surface of the layer [100]. Due to the annealing process, an angle of this layer increased after the deposition of SWCNTs. However, the substrate retained the hydrophilicity with an angle of 57.9°. On the other hand, the e-PVDF-HFP layer was hydrophobic with an angle of 95.9°, and this was reported to be due to low surface energy that was developed by fluorine atoms in the FKM. As can be seen in **Figure 20(a)**, the hydrophobicity of e-PVDF-HFP layer can therefore be a permanent barrier to moisture, water/oxygen molecules to hamper the physisorption on the surface of SWCNTs and the oxide layer [100].

Hiao et al. [85] prepared, as shown in **Figure 21**, the porous conductive fluororubber-SWCNTs composites that were said have great stability for pressure sensing applications in electronics sector. They homogenized the pristine SWCNTs with a foaming agent called N,N-dinitrosopentamethylenetetramine (DPT) into 20 ml of methyl ethyl ketone (MEK) using a magnetic stirrer. Dispersion of SWCNTs in MEK was ensured by further sonicating and homogenizing (at 200 W for 10 min) the SWCNTs/DPT mixture, which was then mixed with about 3 g of Daiel-G801 fluororubber. The resultant SWCNTs/G801 suspension (fluororubber-SWCNTs composite) was stirred for 5 h with the aim of dissolving the fluororubber. After

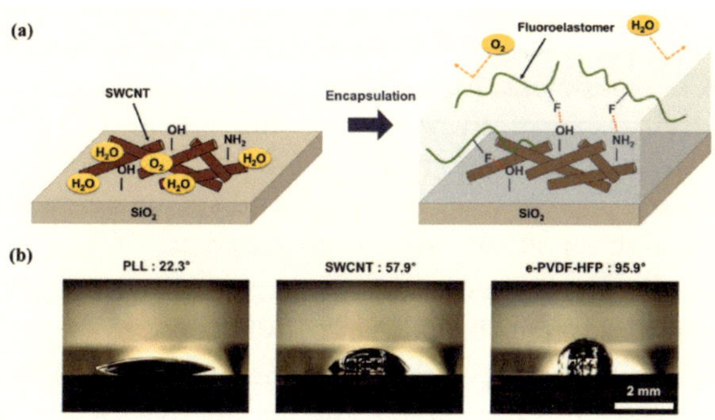

Figure 20.
(a) Effect of encapsulating a layer of SWCNTs-based thin-film transistors. (b) Optical images of contact angle measurements on different layer surfaces [100].

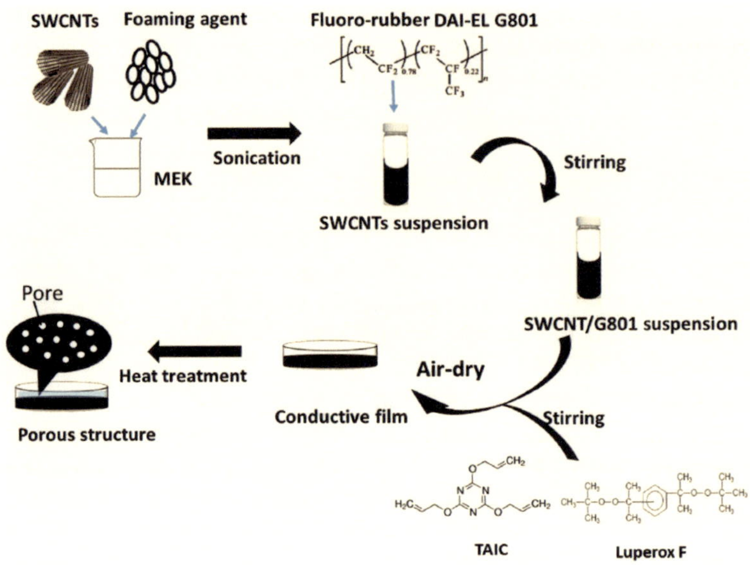

Figure 21.
Preparation of porous conductive fluororubber-MWCNTs composites [85].

being air-dried at ambient temperature for 24 h, the fluororubber-SWCNTs mixture, containing the vulcanizing agents (0.15 g of Luperox F and 0.39 g of triallyl isocyanuric acid), was vulcanized on a hot press machine at about 160°C for 15 min and 180°C for 1 h. The foaming process was initiated by increasing the temperature of the press to about 210°C to obtain a porous structure after about 1 h. To print the conductive silver plates for sensing electrodes (0.5×0.5 cm^2) on polyimide films, a dispenser was employed. For 30 min, a sensor of the same size, pressed between the two printed electrodes, was baked at about 200°C. The prepared porous pressure-sensitive fluororubber-SWCNTs composites showed the electric resistance variation to compressive stresses with a sensitivity that is as high as 4.31 MPa^{-1} when the foaming agent (i.e., DPT) was incorporated into the composite; therefore showing a great potential for fast assembly into the printed electronic circuits and utilization in dynamic pressure monitoring applications in the electronics sector [85].

The properties of FKM have been seen to improve after the incorporation of MWCNTs [6, 60, 61]. Shajari et al. [61] made stretchable electronics materials, *via* high-shear mixing method, with FKM and CNTs. The authors observed the high electrically conductive network, with corresponding ultralow percolation thresholds of about 0.45 phr and 1.40 phr CNTs content. These results indicated that the prepared FKM-CNTs composites have a wide range of strain sensitivity. At the first strain conductivity plateau, FKM-CNTs composites gave high sensitivity with a gauge factor of 1010 at about 23% strain for 0.6 phr nanotubes content, and of 6750 at 34% strain for 1 phr content. At the second strain conductivity plateau, the composites gave high sensitivity with high gauge factor of about 4×10^4 at about 78% strain for 1.5 phr content, and the composite with 2 phr content corresponded to much higher strain of about 100% with gauge factor of 1.3×10^5. The εb of FKM-CNTs composites was found to be as high as 430% and up to about 232% strain sensitivity. These stretchable and conductive FKM-CNTs composite materials are said to be best suited for wearable electronics, including stretchable sensors and light-emitting diodes (LEDs), as well as human motion monitoring electronics [61]. Yang et al. [60] chemically modified the surface of the MWCNTs to get MWCNTs-COOH, MWCNTs-NH_2 and MWCNTs-A1120 filler products, as shown in **Figure 22**. MWCNTs-COOH was prepared by incorporating pristine MWCNTs into a premixed acid solution of H_2SO_4 and HNO_3 with a volume of ratio of 3:1, after which it was ultrasonically stirred for about 2 h at a temperature of 60°C and washed alternately with deionized water and dehydrated ethyl alcohol. The last step involved the filtration with suction and drying of MWCNTs-COOH in a vacuum oven at 80°C for about 12 h. For MWCNTs-NH_2 preparation, MWCNTs-COOH, N-(3-dimethylaminopropyl)-N′-ethylcarbodiimide hydrochloride (EDAC), and 4-Dimethylaminopyridine (DMAP) were ultrasonically mixed for about 30 min with dehydrated ethyl alcohol, after which ethylenediamine (EDA) was added and stirred for about 24 h. The resulted product was washed alternately with deionized water and dehydrated ethyl alcohol to remove excess reagents, and therefore achieving pure MWCNTs-NH_2. The last functionalized MWCNTs product (i.e., MWCNTs-A1120) was prepared by adding a silane coupling agent A-1120 into a premixed solution of deionized water and dehydrated ethyl alcohol at a weight ratio of ¼ with mixing for 1 h, after which pristine MWCNTs ethyl alcohol dispersion was added into the mixed solution, followed by further mixing for about 3 h.

Comparing all prepared functionalized MWCNTs products in **Figure 22**, MWCNTs-A1120 was reported to have a relatively better interfacial interaction with FKM matrix; therefore, it showed uniform distribution within the matrix, and hence the FKM-MWCNTs-A1120 composite had relatively the best tensile properties (UTS increased by 16.58% compared to that of neat FKM; the εb was maintained above 111% with 0.5 wt.% MWCNTs-A1120 content). The nature of dispersion of MWCNTs-A1120 into FKM matrix is shown by fracture surfaces of FKM-MWCNTs-A1120 composites in **Figure 23**. At 1 wt.% and 3 wt.% MWCNTs-A1120 contents, MWCNTs-A1120 are seen to be uniformly distributed within the FKM matrix. As the content was increased, MWCNTs-A1120 began to contact each other, forming a conductive network. A dramatic agglomeration began to take place at a loading of 7 wt.% MWCNTs-A1120, which can be seen in **Figure 23(d)** (red dashed frames).

Uniform distribution of MWCNTs-A1120 within FKM matrix is attributed to the carbon–carbon double bond (–C = C–), formed in FKM molecular chain during vulcanization process, that chemically bonded to the amino group on MWCNTs-A1120, resulting to a strong interface between MWCNTs-A1120 and FKM matrix [60, 101]. This phenomenon also explains why the composites

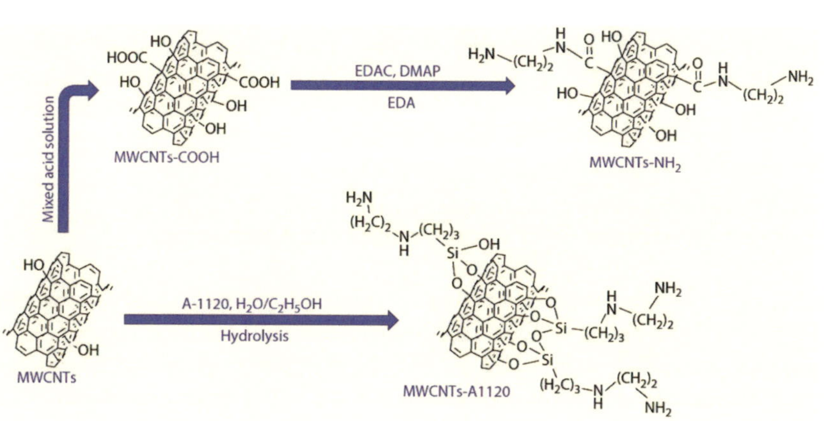

Figure 22.
Schematic representation of surface modification of MWCNTs [60].

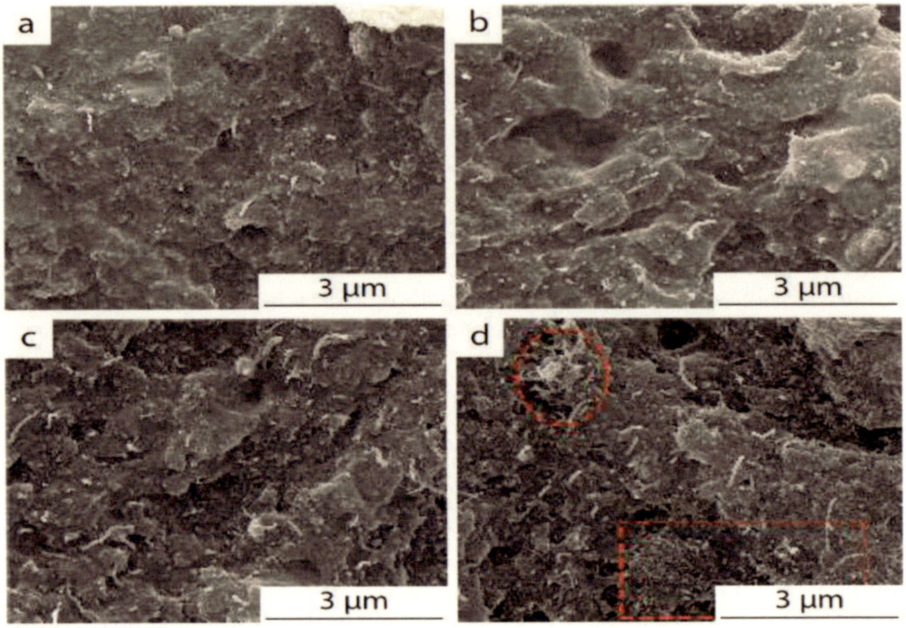

Figure 23.
SEM micrograph of FKM-MWCNTs-A1120 composites: (a) with 1, (b) with 3, (c) with 5 and (d) with 7 wt.% MWCNTs-A1120 content [60].

containing amino MWCNTs (i.e., MWCNTs-NH_2 and MWCNTs-A1120) had outstanding tensile properties [60]. In addition to tensile properties, MWCNTs-A1120 seemed to enhance the electrical properties of FKM-MWCNTs-A1120 composite. As the loading of MWCNTs-A1120 was increased, the dielectric constant and dielectric loss of the FKM-MWCNTs-A1120 composite also increased, and the volume resistivity decreased. When the doping concentration of MWCNTs-A1120 reached 5 wt.%, the dielectric constant and dielectric loss of FKM-MWCNTs-A1120 composite increased significantly, and the volume resistivity got reduced. The authors [60] reported that since a conductive network can be formed by MWCNTs-A1120 product, the doping concentration of 5 wt.% MWCNTs-A1120 can be taken as the percolation threshold of the FKM-MWCNTs-A1120 composite. Nonetheless, as the tensile deformation increased, the dielectric constant and dielectric loss of the composite decreased, and the volume resistivity increased. This may be an indication that the tensile deformation can increase the spacing of the conductive filler or even destroy the conductive network structure, and hence influence the electrical properties of the materials [60]. Therefore, the great potential applications for FKM/MWCNTs-A1120 composites include flexible dielectric and flexible conductive materials [60].

6. Overview of rubber-CNTs composites for electronics

Electronics rubbers generally have some good and poor properties, and their good properties are due to their stable backbone structure and are the main factors that prompted the initial interest for the use of rubber in electronics. For instance, rubbers like NBR, EPDM and FKM are known to have high lifetime or life expectancy, with FKM having relatively the highest, as it has been shown by their hardness and compression set properties which were studied on their O-ring seals after aging for five-years at various temperatures [14]. With the current transition from CB to CNTs due to superior properties and performances of the latter relative to the former, literature has reported that rubber-CNTs composites are currently the major application area for CNTs in the reinforcement of rubbers. Since rubber materials for electronic components need to be precisely shaped to provide proper insulation, electric shock absorption, chemical and thermal resistance and high mechanical strength, and yet be easy to use in the manufacturing and installation of the end product, several researchers have reported that CNTs are the future reinforcing materials for rubbers used in electronics, in partial or complete replacement of conventional fillers, due to the fact that rubber compounds containing CNTs have relatively ultralight weight, extremely high flexibility, and superior chemical, thermal and mechanical properties [8, 2, 35, 49, 102]. Owing to their extremely large surface area, CNTs are expected to enhance the matrices of the electronics rubbers and thereby enhancing their poor properties and further improving their good properties to produce rubber-CNTs composite materials with excellent properties [2, 34, 36, 49].

7. Conclusions

It is notable that rubber-CNTs composites present an array of possibilities for their use in electronics industry. Rubbers such as EPDM, Q, SBR, NBR, FKM/FPM, IR and neoprene are commonly used and often reinforced with conventional fillers to strengthen, insulate and seal the electronic components, including electronic circuits, wires and cables; and thereby protecting them from exposure to high-stress and extreme environments, which normally causes problems that may lead to catastrophic breakdown of the electronic machines and instruments [5–7, 9, 16]. Although the present study found that carbon black (CB) is widely used to reinforce electronics rubber materials, there is also some evidence on the use of CNTs as rubber reinforcing materials, even for rubber materials used in the electronics industry.

The studies about reinforcement of electronics rubbers using CNTs showed that even though CNTs are used only in minute quantities compared to CB, they result in outstanding properties and performances of composites. Several researchers strongly believe that rubber-CNTs composites are futuristic materials mainly for the electrical and thermal insulation since CNTs (both SWCNTs and MWCNTs) can form a conducting network within the composite material at contents that are above the certain minimum value called the percolation threshold.

The properties and performances of the rubber-SWCNTs and rubber-MWCNTs composites are often dependent on the rubber reinforcement quality which is defined by the extent of dispersion of CNTs in the rubber matrix, level of CNTs wetting with rubber, and degree of interfacial bonding between CNTs and the rubber matrix [8, 13, 21, 38]. Surface modification of CNTs and optimization of the ratio of CNTs to the amount of rubber seems to be the main factors that can possibly lead to high exploitation of the properties and performances of rubber-CNTs materials in the electronics application. The suitability of rubber-CNTs composites for the manufacturing of electronics rubber materials is based on that their properties meet most of the specifications for these materials, i.e., rubber-CNTs composites are flexible and light in weight for easy installation and are capable of resisting prolonged vibrations, high strain/pressure and most substances that may cause rubber fatigue and cracks, and therefore expose the electronics to damage [6, 7, 2, 16, 45]. Additionally, rubber compounds reinforced with CNTs can be made to be relatively cost-effective for the manufacturing and maintenance of the CNTs reinforced electronics rubber materials [5, 39, 87]. The performance of SWCNTs is comparable to that of MWCNTs in both the functional and structural composites. However, rubbers like Q are typically expensive, and therefore their nanocomposites often constitute MWCNTs rather than SWCNTs because the former is relatively much cheaper to fabricate than the latter. The preference of conductive SWCNTs filler over MWCNTs by some researchers is due to that the former can result in relatively better homogeneity during the fabrication of rubber composites especially for pressure-sensitive rubber materials. For conductive MWCNTs, it is suggested that the MWCNTs are functionalized before they are incorporated into the rubber matrix, especially for flexible dielectric and flexible conductive materials.

Currently, the preparation of rubber-CNTs composites seems to be dominantly done at the laboratory scale. Therefore, when scaling up for mass production for CNTs reinforced electronics rubber materials in the future, the main challenge that the researchers could potentially still face is to come up with practical ways of overcoming the parameters that typically affect the properties and performance of CNTs in the prepared rubber-CNTs composites.

Acknowledgements

Funding support from the National Research Foundation and Nelson Mandela University is appreciated.

Conflict of interest

The authors declare no conflict of interest.

Author details

Jabulani I. Gumede, James Carson and Shanganyane P. Hlangothi*
Department of Chemistry, Centre for Rubber Science and Technology, Nelson Mandela University, Port Elizabeth, South Africa

*Address all correspondence to: percy.hlangothi@mandela.ac.za

References

[1] Kausar A, Rafique I, Muhammad B. Review of applications of polymer/carbon nanotubes and epoxy/CNT composites. Polymer-Plastics Technology and Engineering. 2016;55(11):1167-1191. DOI:10.1080/03602559.2016.1163588

[2] Abdalla S, Al-Marzouki F, Al-Ghamdi AA, Abdel-Daiem A. Different technical applications of carbon nanotubes. Nanoscale Research Letters. 2015;10(358):1-10. DOI:10.1186/s11671-015-1056-3

[3] Saifuddin N, Raziah AZ, Junizah AR. Carbon nanotubes: A review on structure and their interaction with proteins. Journal of Chemistry. 2013;1-19. DOI:10.1155/2013/676815

[4] Schuler CA. Electronics principles & applications. 9th ed. New York: McGraw-Hill Education; 2019. 625 p. ISBN:978-0-07-337383-6

[5] Ata S. Features and application of carbon nanotube and rubber composites. International Polymer Science and Technology. 2017; 44(8):33-38. DOI:10.1177/0307174X1704400807

[6] Chougule H, Giese U. Application of carbon nanotubes in specialty rubbers – potential and properties. Raw Materials and Applications. 2016;1-8.

[7] Izadi M, Danafar F, Ab Kadir MZA. Natural rubber — Carbon nanotubes composites, recent advances and challenges for electrical applications. In: Proceedings of the IEEE International Conference on Automatic Control and Intelligent Systems (I2CACIS); 2016; Selangor. p. 61-65.

[8] Khalid M, Ratnam CT, Walvekar R, Ketabchi MR, Hoque ME. Reinforced natural rubber nanocomposites: Next generation advanced material. In: Jawaid M, et al., editors. Green Biocomposites, Green Energy and Technology. Springer International Publishing AG; 2017. p. 1-38. DOI:10.1007/978-3-319-49382-4_14.

[9] Ponnamma D, Sadasivuni KK, Wan C, Varughese KT, Thomas S, Alma'adeed MA. Flexible and stretchable electronic composites. Springer Series on Polymer and Composite Materials; 2016. 400 p. DOI:10.1007/978-3-319-23663-6

[10] Nutzenadel C, Zuttel A, Schlapbach CD. Electrochemical storage of hydrogen in nanotube materials. Electrochem Solid State Lett. 1999;2:30-32.

[11] Saint C, Saint JL. Integrated circuit [internet]. Encyclopedia Britannica, inc. 2019. Available from: https://www.britannica.com/technology/integrated-circuit/Bipolar-transistors.

[12] Biswas S, Mozafari M, Reiprich J, Schlag L, Isaac NA, Stauden T, Pezoldt J, Jacobs HO. Metamorphic and stretchable electronic systems – A materials, assembly, and interconnection challenge. Results and Portfolios of Research Institutions. 2018;1-3.

[13] Dai L, Sun J. Mechanical properties of carbon nanotubes-polymer composites. In: Beber MR, Hafez IH, editors. Carbon nanotubes - Current progress of their polymer composites. IntechOpen; 2016. p. 1-41. DOI:10.5772/62635

[14] Kömmling A, Jaunich M, Goral M, Wolff D. Insights for lifetime predictions of O-ring seals from five-year long-term aging tests. Polymer Degradation and Stability. 2020;179:109278. DOI:10.1016/j.polymdegradstab. 2020.109278

[15] Li Z-W. Thermoelectric properties of carbon nanotube/silicone rubber

composites. Journal of Experimental Nanoscience. 2017;12(1):188-196. DOI:10.1080/17458080.2017.1295475

[16] Shen J, Yao Y, Liu Y, Leng J. Preparation and characterization of CNT films/silicone rubber composite with improved microwave absorption performance. Mater. Res. Express. 2019;6:1-13. DOI:10.1088/2053-1591/ab168e

[17] Wang G, Yu Q, Hu Y, Zhao G, Chen J, Li H, Jiang N, Hu D, Xu Y, Zhu Y, Nasibulin AG. Influence of the filler dimensionality on the electrical, mechanical and electromagnetic shielding properties of isoprene rubber-based flexible conductive composites. Composites Communications. 2020;21:100417. DOI:10.1016/j.coco.2020.100417

[18] Boonbumrung A, Sae-oui P, Sirisinha C. Reinforcement of multiwalled carbon nanotube in nitrile rubber: In comparison with carbon black, conductive carbon black, and precipitated silica. Journal of Nanomaterials. 2016;1-9. DOI:10.1155/2016/6391572

[19] Fu DH, Zhan YH, Yan N, Xia HS. A comparative investigation on strain induced crystallization for graphene and carbon nanotubes filled natural rubber composites. eXPRESS Polymer Letters. 2015;9(7):597-607. DOI:10.3144/expresspolymlett.2015.56

[20] Lorenz H, Fritzsche J, Das A, Stöckelhuber KW, Jurk R, Heinrich G, Klüppel M. Advanced elastomer nanocomposites based on CNT-hybrid filler systems. Composites Science and Technology. 2009;69(13):2135-2143. DOI:10.1016/j.compscitech.2009.05.014

[21] Khan W, Sharma R, Saini P. Carbon nanotube-based polymer composites: synthesis, properties and applications. In: Berber MR, Hafez IH, editors. Carbon nanotubes - Current progress of their polymer composites. IntechOpen; 2016. p. 1-47. DOI: 10.5772/62497

[22] Mensah B, Kim HG, Lee J-W, Arepalli S, Nah C. Carbon nanotube-reinforced elastomeric nanocomposites: a review. International Journal of Smart and Nano Materials. 2015;6(4):211-238. DOI:10.1080/19475411.2015.1121632

[23] Fu J-F, Chen L-Y, Yang H, Zhong Q-D, Shi L-Y, Deng W, Dong X, Chen Y, Zhao G-Z. Mechanical properties, chemical and aging resistance of natural rubber filled with nano-Al_2O_3. Polymer Composites. 2012;1-9. DOI: 10.1002/pc.22162

[24] Gao J, He Y, Gong X, Xu J. The role of carbon nanotubes in promoting the properties of carbon black-filled natural rubber/butadiene rubber composites. Results in Physics. 2017;7:4352-4358. DOI:10.1016/j.rinp.2017.09.044

[25] Geckerler KE, Nishide H. Advanced nanomaterials. Weinheim: WILEY-VCH Verlag GmbH & Co, KGaA; 2010:1. 954 p. DOI:10.1002/9783527628940

[26] Al-Saleh MH, Sundararaj U. Electromagnetic interference shielding mechanisms of CNT-polymer composites. Carbon. 2009;47(7):1738-1746. DOI:10.1016/j.carbon.2009.02.030

[27] Chen J, Liu B, Gao X, Xu D. A review of the interfacial characteristics of polymer nanocomposites containing carbon nanotubes. RSC Adv. The Royal Society of Chemistry. 2018;8:28048-28085. DOI:10.1039/c8ra04205e

[28] Chen J, Yan L. Recent advances in carbon nanotube-polymer composites. Advances in Materials. 2017;6(6):129-148. DOI:10.11648/j.am.20170606.14

[29] Hayashida K, Matsuoka Y. Electromagnetic interference shielding properties of polymer-grafted carbon nanotube composites with high electrical resistance. Carbon.

2015;85:363-371. DOI:10.1016/j.carbon.2015.01.006

[30] Kwon S, Ma R, Kim U, Choi HR, Baik S. Flexible electromagnetic interference shields made of silver flakes, carbon nanotubes and nitrile butadiene rubber. Carbon. 2014;68:118-124. DOI:10.1016/j.carbon.2013.10.070

[31] Zhang C-S, Ni, Q-Q, Fu S-Y, Kurashiki K. Electromagnetic interference shielding effect of nanocomposites with carbon nanotube and shape memory polymer. Composites Science and Technology. 2007;67(14):2973-2980. DOI:10.1016/j.compscitech.2007.05.011

[32] Thaptong P, Sirisinha C, Thepsuwan U, Sae-Oui P. Properties of natural rubber reinforced by carbon black-based hybrid fillers. Polymer-Plastics Technology and Engineering. 2014;53:818-823. DOI:10.1080/03602559.2014.886047

[33] Jose T, Moni G, Salini S, Raju AJ, George JJ, George SC. Multifunctional multi-walled carbon nanotube reinforced natural rubber nanocomposites. Industrial Crops & Products. 2017;105:63-73. DOI:10.1016/j.indcrop.2017.04.047

[34] Wayne EJ, Chiguma J, Johnson E, Pachamuthu A, Santos D. Electrically and thermally conducting nanocomposites for electronic applications. Materials. 2010;3:1478-1496. DOI:10.3390/ma3021478

[35] Nah C, Lim JY, Sengupta R, Cho BH, Gent AN. Slipping of carbon nanotubes in a rubber matrix. Polym Int. 2010;1-3. DOI: 10.1002/pi.2909

[36] Kim JH, Hwang J-Y, Hwang HR, Kim HS, Lee JH, Seo J-W, Shin US, Lee SH. Simple and cost-effective method of highly conductive and elastic carbon nanotube/polydimethylsiloxane composite for wearable electronics. Scientific Reports. 2018;8(1375):1-11. DOI:10.1038/s41598-017-18209-w

[37] Iijima S. Helical microtubules of graphitic carbon. Nature. 1991;354:56-58. DOI:10.1038/354056a0

[38] Pravin J, Khan AA, Massimo R, Carlo R, Alberto T. Multiwalled carbon nanotube – strength to polymer composite. Walter de Gruyter GmbH, Berlin/Boston, Physical Sciences Reviews. 2016;1-15.

[39] Bokobza L. Multiwall carbon nanotube elastomeric composites: A review. Polymer. 2007;48:4907-4920. DOI:10.1016/j.polymer.2007.06.046

[40] Kummerlöwe C, Vennemann N, Pieper S, Siebert A, Nakaramontri. Preparation and properties of carbon-nanotube composites with natural rubber and epoxidized natural rubber. Polimery. 2014;59(11-12):1-8. DOI:10.14314/polimery.2014.811

[41] Boonmahitthisud A, Chuayjuljit, S. NR/XSBR nanocomposites with carbon black and carbon nanotube prepared by latex compounding. J. Met. Mater. Miner. 2012;22(1):77-85.

[42] Jogi BF, Sawant M, Kulkarni M, Brahmankar PK. Dispersion and performance properties of carbon nanotubes (CNTs) based polymer composites: A review. J. Encapsulation Adsorpt. Sci. 2012;2:1-10.

[43] Kaushik BK, Majumder MK. Carbon nanotube: properties and applications. SpringerBriefs Appl. Sci. Technol. 2015;1-22.

[44] Donnet J, Voet A. Carbon Black—Physics, Chemistry, and Elastomer Reinforcement. Marcel Dekker: New York. 1976.

[45] Schaefer RJ. Mechanical properties. In: Piersol AG, editor. Harris' shock and vibration handbook. The McGraw-Hill Companies, Inc; 2002. p. 1-18.

[46] Anoop AK, Sunil JT, Rosamma A, Rani J. Natural rubber-carbon nanotube composites through latex compounding. International Journal of Polymeric Materials. 2009;59(1):33-44. DOI:10.1080/00914030903172916

[47] Gumede JI, Carson J, Hlangothi SP, Bolo LL. Effect of single-walled carbon nanotubes on the cure and mechanical properties of reclaimed rubber/ natural rubber blends. Materials Today Communications. 2020;23:1-5. DOI:10.1016/j.mtcomm.2019.100852

[48] Sethi J. Carbon nanotubes filled polyurethane nanocomposites: A filler morphology and surfactant study [thesis]. Tampere: Tampere University of Technology. 2014.

[49] Kim H-J, Sim K, Thukral A, Yu C. Rubbery electronics and sensors from intrinsically stretchable elastomeric composites of semiconductors and conductors. Sci. Adv. 2017;3:1-9. Kim H-J, Sim K, Thukral A, Yu C. Rubbery electronics and sensors from intrinsically stretchable elastomeric composites of semiconductors and conductors. Sci. Adv. 2017;3:1-9.

[50] Kim TA, Kim HS, Lee SS, Park M. Single-walled carbon nanotube/silicone rubber composites for compliant electrodes. Carbon. 2012;50:444-449. DOI:10.1016/j.carbon.2011.08.070

[51] Azam MA, Talib E, Mohamad N, Kasim MS, Rashid MWA. Mechanical and thermal properties of single-walled carbon nanotube filled epoxidized natural rubber nanocomposite Journal of Applied Sciences. 2014;14:2183-2188. DOI:10.3923/jas.2014.2183.2188

[52] Sandler J, Shaffer MSP, Prasse T, Bauhofer W, Schulte K, Windle AH. Development of a dispersion process for carbon nanotubes in an epoxy matrix and the resulting electrical properties. Polymer. 1999;40(21):5967-5971. DOI:10.1016/S0032-3861(99)00166-4

[53] Vijayalekshmi V. Development and characterisation of EPDM/Silicone rubber nanocomposites for high voltage insulators [thesis]. Vandalur, Chennai: B.SAbdur Rahman University; 2013.

[54] Sui G, Zhong WH, Yang XP, Yu YH, Zhao SH. Preparation and properties of natural rubber composites reinforced with pretreated carbon nanotubes. Polymers for Advanced Technologies. 2008;19(11):1543-1549. DOI:10.1002/ pat.1163

[55] Thostenson ET, Li C, Chou TW. Nanocomposites in context. Compos Sci Technol. 2005;65(3-4):491-516. DOI:10.1016/j.compscitech.2004.11.003

[56] Jia ZJ, Wang Z, Xu C, Liang J, Wei BQ, Wu D, Zhu S. Study on poly(methyl methacrylate)/carbon nanotube composites. Mater. Sci. Engineer: A. 1999;271(1-2):395-400. DOI:10.1016/S0921-5093(99)00263-4

[57] Tobias G, Mendoza E, Ballesteros B. Functionalization of carbon nanotubes. In: Bhushan B, editors. Encyclopedia of Nanotechnology. Springer, Dordrecht; 2012. DOI:10.1007/978-90-481-9751-4_48

[58] Kumar V, Lee D-J. Effects of purity in single-wall carbon nanotubes into rubber nanocomposites. Chemical Physics Letters. 2019;715:195-203. DOI:10.1016/j.cplett.2018.11.042

[59] Sui X-M, Giordani S, Prato M, Wagner HD. Effect of carbon nanotube surface modification on dispersion and structural properties of electrospun fibers. Applied Physics Letters. 2009;95(23):1-3. DOI:10.1063/1.3272012

[60] Yang G-Y, Tong L-F, Liu X-B. Design and properties of fluoroelastomer composites via incorporation of MWCNTs with varied modification. Chinese J. Polym. Sci. 2020;38:983-992. DOI:10.1007/s10118-020-2405-y

[61] Shajari S, Mahmoodi M, Rajabian M, Karan K, Sundararaj U,

Sudak LJ. Highly sensitive and stretchable carbon nanotubes/Fluoroelastomer nanocomposite with a double-percolated network for wearable electronics. Advanced Electronic Materials. 2020;6(2):1901067. DOI:10.1002/aelm.201901067

[62] Grady BP. Carbon nanotube-polymer composites: manufacture, properties, and applications. Hoboken: John Wiley & Sons, Inc; 2011. p. 352. ISBN:978-0-470-59641-8

[63] Wei C, Srivastava D, Cho K. Thermal expansion and diffusion coefficients of carbon nanotube-polymer composites. Nano Letters. 2002;2(6):647-650. DOI:10.1021/nl025554+

[64] Wang C, Xia K, Wang H, Liang X, Yin Z, Zhang Y. Advanced carbon for flexible and wearable electronics. Advanced Materials. 2019;31(9):1801072. DOI:10.1002/adma.201801072

[65] Li Y, Shimizu H. Toward a stretchable, elastic, and electrically conductive nanocomposite: Morphology and properties of poly[styrene-*b*-(ethylene-*co*-butylene)-*b*-styrene]/Multiwalled carbon nanotube composites fabricated by high-shear processing. Macromolecules. 2009;42(7):2587-2593. DOI:10.1021/ma802662c

[66] Sekitani T, Someya T. Stretchable, large-area organic electronics. Adv. Mater. 2010;22(20):2228-2246. DOI:10.1002/adma.200904054

[67] Felhös D, Karger-Kocsis J, Xu D. Tribological testing of peroxide cured HNBR with different MWCNT and silica contents under dry sliding and rolling conditions against steel. J. Appl. Polym. Sci. 2008;108(5):2840-2851. DOI:10.1002/app.27624

[68] Kolodziej M, Bokobza L, Bruneel JL. Investigations on natural rubber filled with multiwall carbon nanotubes. Compos Interfaces. 2007;14(3):215-228. DOI:10.1163/156855407780340304

[69] Bizhani H, Katbab AA, Lopez-Hernandez E, Miranda JM, Verdejo R. Highly deformable porous electromagnetic wave absorber based on ethylene–propylene–diene monomer/multiwall carbon nanotube nanocomposites. Polymers. 2020;12(4),858:1-14. DOI:10.3390/polym12040858

[70] Kang I, Khaleque MdA, Yoo Y, Yoon PJ, Kim S-Y, Lim, KT. Preparation and properties of ethylene propylene diene rubber/multi walled carbon nanotube composites for strain sensitive materials. Composites Part A: Applied Science and Manufacturing. 2011;42(6):623-630. DOI:10.1016/j.compositesa.2011.01.021

[71] Molavi FK, Soltani S, Naderi G, Bagheri R. Effect of multi-walled carbon nanotube on mechanical and rheological properties of silane modified EPDM rubber. Polyolefins Journal. 2016;3(2):69-77. DOI:10.22063/POJ.2016.1293

[72] Shang S, Gan L, Yuen MC, Jiang S, Luo NM. Carbon nanotubes based high temperature vulcanized silicone rubber nanocomposite with excellent elasticity and electrical properties. Composites: Part A. 2014;66:135-141. DOI:10.1016/j.compositesa.2014.07.014

[73] Katihabwa A, Wang W, Jiang Y, Zhao X, Lu Y, Zhang L. Multi-walled carbon nanotubes/silicone rubber nanocomposites prepared by high shear mechanical mixing. Journal of Reinforced Plastics and Composites. 2011;30(12):1007-1014. DOI:10.1177/0731684410394008

[74] Wang P, Geng S, Ding T. Effects of carboxyl radical on electrical resistance of multi-walled carbon nanotube filled silicone rubber composite

under pressure. Compos Sci Technol. 2010;70:1571-1573. DOI:10.1016/j.compscitech.2010.05.008

[75] Bannych A, Katz S, Barkay Z, Lachman N. Preserving softness and elastic recovery in silicone-based stretchable electrodes using carbon nanotubes. Polymers. 2020;12:1345. DOI:10.3390/polym12061345

[76] De Falco A, Goyanes S, Rubiolo GH, Mondragon I, Marzocca A. Carbon nanotubes as reinforcement of styrene-butadiene rubber. Appl Surf Sci. 2007;254(1):262-265. DOI:10.1016/j.apsusc.2007.07.049

[77] Girun N, Ahmadun F-R, Rashid SA, Atieh MA. Multi-wall carbon nanotubes/styrene butadiene rubber (SBR) nanocomposite. Fullerenes, Nanotubes, and Carbon Nanostructures. 2007;15:207-214. DOI:10.1080/15363830701236449

[78] Liu Z, Qian Z, Song J, Zhang Y. Conducting and stretchable composites using sandwiched graphene-carbon nanotube hybrids and styrene-butadiene rubber. Carbon. 2019;149:181-189. DOI:10.1016/j.carbon.2019.04.037

[79] Shao C, Wang Q, Mao Y, Li Q, Wu C. Influence of carbon nanotubes content on the properties of acrylonitrile-butadiene rubber/cobalt chloride composites. Materials Research Express. 2019;6(7):075323. DOI:10.1088/2053-1591/ab1445

[80] Valentini L, Bon SB, Hernandez M, Lopez-Manchado MA, Pugno NM. Nitrile butadiene rubber composites reinforced with reduced graphene oxide and carbon nanotubes show superior mechanical, electrical and icephobic properties. Composites Science and Technology. 2018;166:109-114. DOI:10.1016/j.compscitech.2018.01.050

[81] Verge P, Peeterbroeck S, Bonnaud L, Dubois P. Investigation on the dispersion of carbon nanotubes in nitrile butadiene rubber: Role of polymer-to-filler grafting reaction. Composites Science and Technology. 2010;70(10):1453-1459. DOI:10.1016/j.compscitech.2010.04.022

[82] Wang X, Chen D, Zhong W, Zhang L, Fan X, Cai Z, Zhu M. Experimental and theoretical evaluations of the interfacial interaction between carbon nanotubes and carboxylated butadiene nitrile rubber: Mechanical and damping properties. Materials and Design. 2020;186:108318. DOI:10.1016/j.matdes.2019.108318

[83] Bernal-Ortega P, Bernal MM, Gonzalez-Jimenez A, Posadas P, Navarro R, Valentin JL. New insight into structure-property relationships of natural rubber and styrene-butadiene rubber nanocomposites filled with MWCNT. Polymer. 2020;201:122604. DOI:10.1016/j.polymer.2020.122604

[84] Song SH. The effect of clay/multiwall carbon nanotube hybrid fillers on the properties of elastomer nanocomposites. International Journal of Polymer Science. 2018;1-9. DOI:10.1155/2018/5295973

[85] Hsiao F-R, Wu I-F, Liao Y-C. Porous CNT/rubber composite for resistive pressure sensor. Journal of the Taiwan Institute of Chemical Engineers. 2019;102:387-393. DOI:10.1016/j.jtice.2019.05.017

[86] Kummerlöwe C, Nakaramontri Y, Nakason C, Vennemann N. Simple, efficient preparation of high-performance natural rubber composites. Soc. Plastics Eng: Plastics Res. Online. 2014;1-3. DOI:10.2417/spepro.005629

[87] Kang I, Heung YY, Kim JH, Lee JW, Gollapudi R, Subramaniam S, Narasimhadevara S, Hurd D, Kirikera GR, Shanov V, Schulz MJ, Shi D, Boerio J, Mall S, Ruggles-Wren M. Introduction to carbon nanotube and nanofiber smart materials. Composites Part B:

Engineering. 2006;37(6):382-394. DOI:10.1016/j.compsitesb.2006.02.011

[88] Junian S, Makmud MZH, Sahari J. Natural rubber as electrical insulator: A review. Journal of Advanced Review on Scientific Research. 2015;6(1):28-42.

[89] Kaltseie R, Keplinger C, Koh SJA, Baumgartner R, Goh YF, Ng WH, Kogler A, Tröls A, Foo CC, Suo Z, Bauer S. Natural rubber for sustainable high-power electrical energy generation. RSC Adv. 2014;1-11. DOI:10.1039/x0xx00000x

[90] Hanhi K, Poikelispää M, Tirilä H-M. Elastomeric materials. Plastics and Elastomer Technology (VERT). 2007;1-84.

[91] Haj-Ali R, Zemer H, El-Hajjar R, Aboudi J. Piezoresistive fiber-reinforced composites: A coupled nonlinear micromechanical–microelectrical modeling approach. International Journal of Solids and Structures. 2014;51:491-503. DOI:10.1016/j.ijsolstr.2013.10.022

[92] Shit SC, Shah P. A review on silicone rubber. Natl. Acad. Sci. Lett. 2013;36(4):355-365. DOI:10.1007/s40009-013-0150-2

[93] Momen G, Farzaneh M. Survey of micro/nano filler use to improve silicone rubber for outdoor insulators. Rev.Adv. Mater.Sci. 2011;27:1-13.

[94] Kumar V, Lee G, Monika, Choi J, Lee D-J. Studies on composites based on HTV and RTV silicone rubber and carbon nanotubes for sensors and actuators. Polymer. 2020;190:122221. DOI:10.1016/j.polmer.2020.122221

[95] Lin J, Su S, He Y, Kang F. Improving thermal and mechanical properties of the alumina filled silicone rubber composite by incorporating carbon nanotubes. New Carbon Materials. 2020;35(1):66-72. DOI:1016/S1872-5805(20)60476-0

[96] Yanagizawa A. Tribology Kaigi. Spring, E28; 2015.

[97] Liu Y, Zong H, Gao S, Du BX. Contamination deposition and discharge characteristics of outdoor insulators in fog-haze conditions. Electrical Power and Energy Systems. 2020;121:106176. DOI:10.1016/j.ijepes.2020.106176

[98] Do TN, Visell Y. Stretchable, twisted conductive microtubules for wearable computing, robotics, electronics, and healthcare. Scientific Reports. 2017;7(1753):1-12. DOI:10.1038/s41598-017-01898-8

[99] Ning N, Ma Q, Liu S, Tian M, Zhang L, Nishi T. Tailoring dielectric and actuated properties of elastomer composites by bioinspired poly(dopamine) encapsulated graphene oxide. ACS Appl. Mater. Interfaces. 2015;7(20):10755-10762. DOI:10.1021/acsami.5b00808

[100] Seo J, Ha J, Lee B, Kim H, Hong Y. Fluoroelastomer encapsulation for enhanced reliability of solution processed carbon nanotube thin-film transistors. Thin Solid Films. 2020;704:138021. DOI:10.1016/j.tsf.2020.138021

[101] Gao W, Guo J, Xiong J, Smith AT, Sun L. Improving thermal, electrical and mechanical properties of fluoroelastomer/amino-functionalized multi-walled carbon nanotube composites by constructing dual crosslinking networks. Compos. Sci. Technol. 2018;162:49-57. DOI:10.1016/j.compscitech.2018.04.022

[102] Aryasomayajula L, Wolter K-J. Carbon nanotube composites for electronic packaging applications: A review. Journal of Nanotechnology. 2013;1-7. DOI:10.1155/2013/296517

Section 4

Carbon Nanotube Based Materials and Devices for Energy Storage Application

Chapter 6

Carbon Nanotubes: Applications to Energy Storage Devices

Ruhul Amin, Petla Ramesh Kumar and Ilias Belharouak

Abstract

Carbon nanotubes (CNTs) are an extraordinary discovery in the area of science and technology. Engineering them properly holds the promise of opening new avenues for future development of many other materials for diverse applications. Carbon nanotubes have open structure and enriched chirality, which enable improvements the properties and performances of other materials when CNTs are incorporated in them. Energy storage systems have been using carbon nanotubes either as an additive to improve electronic conductivity of cathode materials or as an active anode component depending upon structural and morphological specifications. Furthermore, they have also been used directly as the electrode material in supercapacitors and fuel cells. Therefore, CNTs demand a huge importance due to their underlying properties and prospective applications in the energy storage research fields. There are different kinds of carbon nanotubes which have been successfully used in batteries, supercapacitors, fuel cells and other energy storage systems. This chapter focuses on the role of CNTs in the different energy storage and conversion systems and impact of their structure and morphology on the electrochemical performances and storage mechanisms.

Keywords: carbon nanotube, energy storage, nanocomposite, batteries, fuel cells, supercapacitor

1. Introduction

Carbon is one of the most important elements on earth and it plays a crucial role in living organisms and modern technological world either as complex compounds or in its elemental form. Carbon has several allotropes (e.g. graphite, diamond, lonsdaleite, Buckyball and amorphous carbon etc.) and different morphological textures (nanotube, nanowire and graphene). Specific applications in devices and other uses are highly specific to the textures and nature of the allotrope of desired properties. Notably, ever since graphite and diamond were discovered for the first time in 1779, their innovative applications have been growing untill the present. Leveraging the benefits of these carbon morphologies, the journey towards innovation and discovery has continued to advance at a steady pace and almost two centuries later, Sumio Iijima discovered for the first time the existence of multi-walled carbon nanotubes (MWCNTs) and in 1992 he observed single-walled CNTs (SWCNTs) [1]. The synthesis and characterization of CNTs is beyond the scope of this chapter. It should be noted that graphite and CNTs have some characteristic properties and features, that enable them to be used in the energy storage and conversion systems. It is worth mentioning that the carbon nanotubes (CNTs), have been envisioned to potentially impact different areas of science and technology due to their unique properties and structural features [2–4]. Specifically, CNTs have very high tensile strength of 60 GPa and high electronic conductivity reported to be 10^8 Scm^{-1} and 10^7 Scm^{-1} for single-walled and multi-walled carbon nanotubes, respectively [5, 6]. Besides the potential practical applications in chemical and bio sensors [7, 8], field emission materials [9], catalyst [10], electronic devices [11], CNTs have been used in energy storage and conversion systems like, alkali metal ion batteries [12], fuel cells [13], nano-electronic devices [14] supercapacitors [15], and hydrogen storage devices [16]. The extraordinarily high electronic conductivity of CNTs enable CNT and graphite as an additive to composite electrodes and facilitate activation of poorly conducting electrode materials making them electrochemically active. In this chapter, we emphasize the applications of CNTs in four different areas: alkali metal ion (Li, Na and K) batteries, alkali metal air batteries, supercapacitors, and fuel cells. The underlying governing structural features and morphological impact on the electrochemical performances have been discussed and the specific storage mechanisms are also highlighted.

2. Structure and properties of carbon nanotubes

Carbon nanotubes can be either as single-walled carbon nanotubes (SWCNTs) or multi-walled carbon nanotubes (MWCNTs). Simply a wrapped graphene sheet with a hallow fiber is the single-walled CNT. On the other hand, a combination and collection of SWCNTs is the multi-walled CNTs. It should be noted that carbon nanotubes are designated as one-dimensional (1D) structures because of the long length-to-diameter ratio (aspect ratio) [17]. The electronic properties of CNTs are associated with the geometrical structure of them which is uniquely specified by a pair of indexes called chiral indexes (n, m). There are three typical types of CNTs can be obtained: armchair (n, n), zigzag (n,0), and chiral (n, m), depending on the orientation of the graphene lattice with respect to the tube axis they are twisted [18–20]. The formation of a single-walled CNT is shown in **Figure 1** by rolling a single graphene sheet in different directions. It is worth to mention that the rolling introduces strain into the carbon bonds oriented circumferentially while the single graphene sheet is made into a tube. This strain will be greater for smaller diameters; therefore, the armchair will be more strained than zigzag single-walled CNTs [21].

Properties of CNTs*:* Importantly, the local electronic character of carbon nanotubes is highly dependent on the carbon framework arrangements either zigzag or armchair. Also, there is a long-range defect which is formed by displacement or disorientation of standard nanocarbon structures, including hybridization of carbons, vacancies creation, and bond rotations (*Stone – Wales)*. These imperfections are responsible for the chemical, mechanical and optoelectronic properties of CNTs. Noting that this imperfection can modify the electronic properties of CNTs by creating Fermi levels variation and the resulted charge diffusion processes can be affected [22, 23].

The resistivity of CNTs resulted from the electrical conductivity is determined by their carbon framework (graphite) and the one-dimensional character which is regulated by the quantum mechanical properties. The resistance of CNTs is independent of the length of the tube and act as a good conductor in which the highest current density can be as high as 10^9 A cm^{-2}. This important property of CNTs may improve the rate capability of electrochemical devices like batteries and capacitors. The helicity and diameter of CNT determines either it would be metallic or semiconducting in nature [24]. It should be mentioned that the strong C=C double bonds in the carbon nanotubes makes them having high Young's modulus in its axial direction and highest tensile strength. Of course, the presence of imperfection/defects in the tube wall reduces the Young's modulus and tensile strength remarkably. However, reported experimental data are significantly smaller than the theoretical predictions which is most probably resulted from the high flexibility and aspect ratio [25, 26]. At room temperature the thermal conductivities of individual SWCNTs is reported up to 6600 W/(m K) which is almost double than the pure diamond [27]. Besides these, the CNTs have many others useful properties such as electro-optic effect, saturable absorption and *Kerr effect* etc. [28, 29].

The favorable and beneficial electrical, mechanical and thermal properties of carbon nanotubes are promising for various electrochemical applications like batteries, supercapacitors, fuel cells and hydrogen storage. Some important properties of SWCNTs and MWCNTs are listed in **Table 1**.

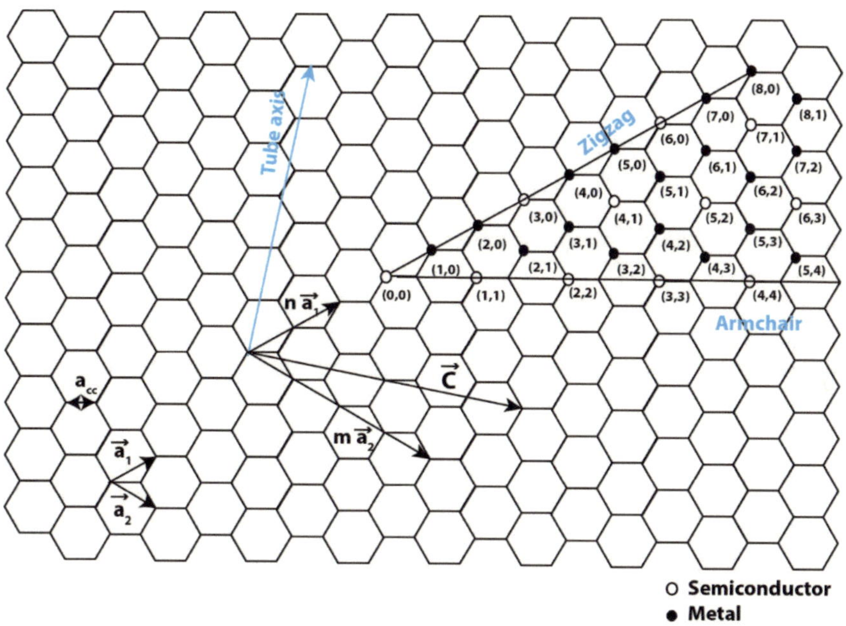

Figure 1.
Lattice, two off-set triangular sublattices of graphene and graphene sheet rolling vector map. Reproduced from Ref. [21] with permission from the Royal Society of Chemistry.

Property	SWCNT	MWCNT
Specific gravity	0.8 g cm^{-3}	<1.8 g cm^{-3}
Elastic modulus	~1.4 TPa	~0.3–1 TPa
Resistivity	5–50 $\mu\Omega$ cm	5–50 $\mu\Omega$ cm
Thermal conductivity	3000 W m^{-1} K^{-1}	3000 W m^{-1} K^{-1}
Magnetic susceptibility	22 × 106 EMU g^{-1}	22 × 106 EMU g^{-1}
Thermal expansion	Negligible	Negligible
Thermal stability	600–800°C (air) 2800°C (vacuum)	600–800°C (air) 2800°C (vacuum)
Strength	50–500 GPa	10–60 GPa

Reproduced from Ref. [30] with permission from the American Chemical Society.

Table 1.
Properties of single walled and multi walled nanotubes.

The values of Young's modulus and tensile strength of CNTs are around 1.2 TPa and 160 GPa, respectively. These unique mechanical properties make CNTs one of the toughest materials and play a vital role in protecting electrode integrity during the charge–discharge cycle of alkali-metal ion batteries. Furthermore, CNT based paper can be used as active material and current collector in supercapacitors, which can reduce the contact resistance as well as electrode weight. The thermal stability of CNTs is also an important property, which can help the composite electrode for stable battery operation at high current rates. SWCNTs and DWCNTs are showing a positive thermal expansion coefficient of $1.9 \times 10^{-5}\ K^{-1}$ and $2.1 \times 10^{-5}\ K^{-1}$, respectively, at room temperature. This negligible thermal expansion coefficient makes CNTs feasible for high energy density battery applications.

3. Storage mechanism of carbon nanotubes in electrochemical applications

CNTs have showed high performance as anode materials and cathode additive for alkali metal ion batteries because of their favorable properties (electrical, mechanical, and structural). The battery electrode based on CNTs attracted attention of many research groups around the world. Recently different modifications in the CNTs have been made for the deployment as a promising electrode material regarding alkali metal ion intercalation, adsorption, and diffusion [31]. In Lithium ion Batteries (LIBs), it has been well established that Li^+ ions are stored via two mechanisms, one is intercalation and other one is alloying [32]. The lithium ion storage mechanism in CNTs have been investigated by many research groups. First, let us go into detail about intercalation mechanism in pure carbon nanotubes. Because of different morphologies, the amount of Li^+ ion insertion is not limited to LiC_6. The capacities (Li ion storage capacity) is highly dependent to the CNT morphology, especially defects and diameter of the carbon nanotubes [33].

Defects (n-rings) can be occured naturally or introduced by treatment (nitric acid treatment or ball milling) as shown **Figure 2**. The theoretical studies (DFT total-energy calculations using local-density approximation (LDA) and the

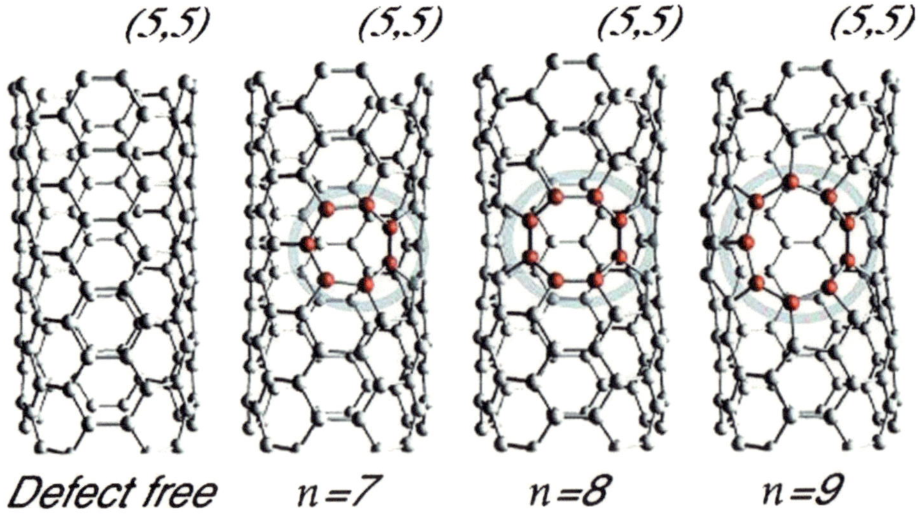

Figure 2.
Types of defects (rings of the red dots) in a (5,5) SWCNT. Reproduced from Ref. [34] with permission from the American Physical Society.

generalized-gradient approximation (GGA)) were employed to investigate the detailed energetics of lithium ion adsorption on the defective single-wall carbon nanotubes [34]. It turned out that the presence of the holes on CNTs wall increase their capacity which is most likely favorable diffusion of lithium ion into the inside of the carbon nanotubes and reduces the diffusion path length [35, 36].

Another important note is that Li^+ ion can also penetrate the CNTs from its ends. Meunier *et al.*, adopted *ab initio* simulations for investigating the lithium ion migration through the ends of open-ended carbon nanotubes [37]. It is obvious that the CNTs should be short in size to allow Li^+ ions to freely intercalate/de-intercalate. The theoretical studies indicated that the capacity difference between opened and closed carbon nanotubes was almost 120 mAh g^{-1} [38]. It is also reported that the reversible Li storage capacity increased from LiC_6 in closed ended tubes to LiC_3 after etching which might be due to the short and highly defective CNTs generation [33]. Once Li^+ ion entered CNT either from the end of tube or through defects, it undergoes one 1D random walk in the tube. Provided that if the tube is very large the Li^+ ion will be able enter, however, it will be difficult to exit or never exit. It is indirectly proved by Wang et al., showing that capacity of a short (300 nm) CNTs is much higher than the longer CNTs (micro-meter) [39]. On the other hand, Yang *et al.*, investigated the impact of length on electrochemical properties of CNTs. It was observed that the small size CNTs exhibit relatively less charge-transfer resistance than longer CNTs. It is not clearly explained why the lithium ion diffusion coefficients (D_{Li}) of both the long and short CNTs reduces as the intercalation is on progress and voltage drops. It might be due to repulsive interactive as lithium concentration increases in the tube. However, in short CNTs the difference between initial and final value of diffusion was smaller than longer CNTs. Therefore, the investigation indicates that shorter the CNTs length better will be the electrochemical performances [36]. In addition, Wang *et al.*, developed solid state cutting method to prepare the short CNTs from micro-meter long CNTS using Nickel Oxide (NiO)

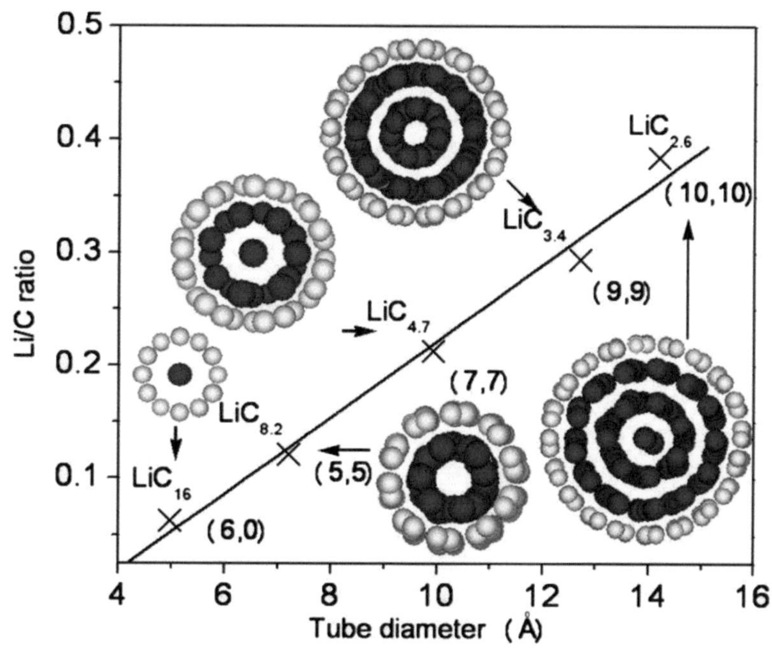

Figure 3.
The variation of Li/C ratio as a function of tube diameter [White and grey balls represent C and Li atoms]. Reproduced from Ref. [41] with permission from Elsevier.

particle as a cutter at 900°C. They successfully obtained short CNTs around 200 nm in length and the measured electrochemical reversible capacities increases as the length of CNT is decreases [40]. The same research group used Iron (II) sulfide (FeS) as a catalyst to produce the short CNTs as well as directly grown short CNTs with length of 200–500 nm. They are able to show that the long CNTs exhibited 188 mAh g^{-1} while short CNTs 502 mAh g^{-1} capacity [39].

Furthermore, there is significant relationship between the ratio of lithium-carbon (Li/C) and the diameter of tube. If the tube diameter is bigger, the intercalated lithium atoms gravitated to form multi-shell structural feature when the system is at the equilibrium state (**Figure 3**). These structures with a linear chain in the axis will improve the lithium capacity. It was also reported that the interaction potential at the central region is varied with the diameter of the nanotubes and diameter of 4.68 Å has higher interaction energy, that made CNTs better candidate for lithium ion battery anode material [41, 42].

Another important factor for lithium storage in CNTs is conducting nature of CNTs. There are two different types of CNTs, as mention above, one is semi-conducting another one is metallic CNTs based on their chirality. The experimental measurements and modeling studies indicated that if the chiral vector is a multiple of 3, the CNT behaves like metallic; otherwise it would be semiconducting. The metallic CNTs is able to store approximately 5 times more lithium ions than semi-conducting CNTs [43].

4. Electrochemical applications

4.1 Carbon nanotubes and their composites for alkali metal ion batteries (Li, Na and K) and other batteries

As it is discussed above, the one-dimensional carbon nanotube can be obtained as single-walled carbon nanotubes and multiwalled carbon nanotubes. Last 20 years, applications of CNTs are emerging in energy storage research on carbon structures and nano composite materials because of their excellent electrochemical properties including lower density, higher tensile strength, and higher rigidity [44].

Li-Ion Batteries (LIBs): Both single walled and multi walled carbon nanotubes are highly investigated in lithium ion battery either as an anode material or as a conductive additive in the composite electrodes. It is worth mentioning here that the one-dimensional CNTs enable to store higher amount of lithium than the conventional graphitic carbon (specific capacity of 372 mAh g^{-1}). The CNTs exhibits reversible capacities range between 300 and 1250 mAh g^{-1}, depending on structure and morphology and defect concentration [44–47]. The SWCNTs show first discharge capacity around 2500 mAh g^{-1} with a voltage plateau between 1 and 2 V vs. Li/Li^{+}. However, after first charge–discharge cycle the voltage profile varies based on the quality of CNTs and their pre-treatment [48]. Yang *et al*., prepared unetched SWCNTs by co-pyrolysis method and the measured capacity was 170 mAh g^{-1} and 266 mAh g^{-1} for differently treated two samples [36] although the theoretical studies indicates that the reversible capacities should be more than 1116 mAh g^{-1} (LiC_2 stoichiometry) as it is possible for single walled CNTs [49].

Along with SWCNTs, researchers successfully demonstrated the lithium ion intercalation into MWCNTs [50] (**Figure 4**). It is interesting to note that the specific capacities around 8500 mAh g^{-1} was reported for multi-walled CNTs at slow current rate (0.1 mA cm^{-2}). On the Contrary, however, most of the carbon nanotubes show capacities typically less than 4000 mAh g^{-1} [44]. A comparative study has been carried out on highly conductive, binder-free, free-standing flexible films

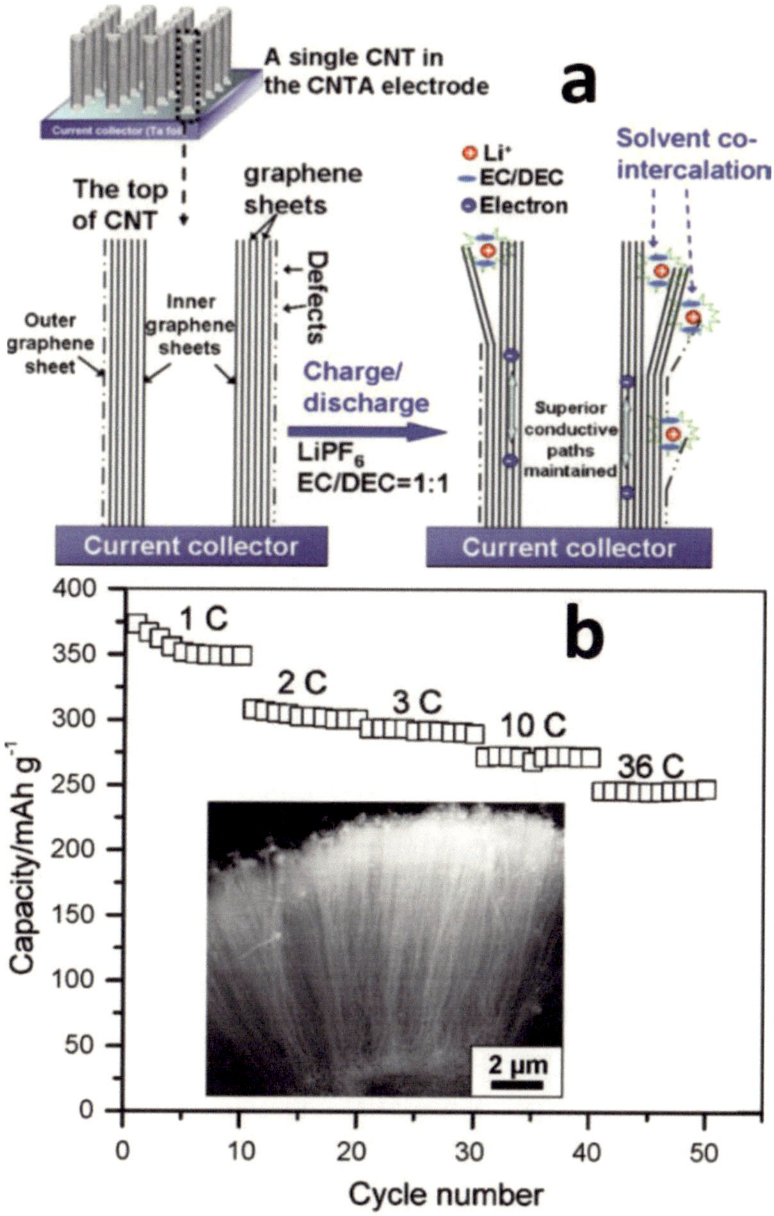

Figure 4.
(a) Schematic representation of the microstructure of nanotube array and energy storage mechanism and (b) cycle performance of carbon nanotube array (CNTA) electrodes. Reproduced from Ref. [50] with permission from Elsevier.

made from three different types of carbon nanotubes (SWCNTs, DWCNTs and MWCNTs). They were able to show that the free standing MWCNT film was retain its capacity after hundreds cycles, which is better than other CNTs films [51]. Lahiri *et al.*, prepared directly deposited MWCNTs on cooper current collector by chemical vapor deposition (CVD). It exhibits better specific capacity, at high current rate of 3C and good cyclic stability over 50 cycles [52].

Charan *et al.*, prepared aligned multiwalled carbon nanotubes (MWNTs) on stainless-steel foil and obtained high stable specific capacity of 460 mAh g^{-1} for 1200 cycles at 1C rate [53]. Li et al., synthesized stacked multiwall carbon nanotubes (MWCNTs) by floating catalyst chemical vapor deposition (FC-CVD) method and observed a stable discharge capacity of 310 mAh g^{-1} at 0.5 C rate for 300 cycles [54]. Brian *et al.*, obtained highest capacity for SWCNT electrodes with using 1 M $LiPF_6$ in Ethylene carbonate: Propylene carbonate: Dimethyl carbonate (EC:PC:DMC) as the electrolyte and capacity retention is more than 95% after 10 cycles [55]. Researchers have using different surface functionalization and doping (N, B) processes for getting efficient Li ion storage in CNTs [56] and highly concentrated N doped CNTs was developed and presented reversible capacity of 494 mAh g^{-1} which is almost double conventional CNTs capacity [57]. On the other hand, when flexible and free-standing pyridin-B-CNTs film was prepared using one-step floating catalyst chemical vapor deposition method, it delivers high specific capacity with excellent cycle stability of 548 mAh g^{-1} after 300 cycles at 0.1 A g^{-1} [56].

Up to now discussion was concentrated on the raw CNTs utilization in lithium ion battery as an anode material. Hereafter the discussion will be focused on the collective data for hybrid nanocomposites by incorporating CNTs into Li-storage compounds as new electrode (anode & cathode) materials. In this composite electrode, significance of π-orbital overlap in metallic type CNTs where electrons can transfer with mean free paths along the length of the nanotube (ballistic transport). So, when it is used as an additive, it will increase rate performance, especially combined with the poor electronic conductive cathode materials. Furthermore, CNTs have the mechanical and electrical properties along with a large surface area which is beneficial for lithium ion battery composite electrode [48]. The CNT was employed in silicon based anode consisting of silicon nanowire/graphene sheet (SiNW@G) which was intertwined architectures [58] where CNT can act either as conductive additive or active component depending on the operation voltage of the cell. The molybdenum dioxide was embedded with multiwalled carbon nanotubes (MoO_2/MWCNT) by hydrothermal process where hybrid composite consists of spherical flowerlike MoO_2 nanostructures interconnected by MWCNTs and exhibits reversible lithium storage capacity of 1143 mAh g^{-1} at a current density of 100 mA g^{-1}. The zinc oxide was covered by N-doped carbon freestanding membrane electrodes for lithium ion batteries and the hybrid material shows the high performance with a specific capacity (850 mAh g^{-1}at a current density of 100 mA g^{-1}) and excellent cycling stability [59]. The polymer-derived silicon oxy-carbide/carbon nanotube (SiOC/CNT) composites exhibit stable lithium anode material [60].

The application of carbon nanotubes as an additive for anode or cathode has huge advantages compared to other carbon form like amorphous carbon, acetylene black tc.. As discussed above the CNTs have a high electrical conductivity at room temperature and very small amount (0.2% w/w) of CNTs will be able to create a percolation network for electronic conductivity [61] and therefore, could increase orders of magnitude in electrical conductivity of composite electrodes and form better percolation network. CNTs have been employed as an conducting additive for $LiCoO_2$, $LiNi_{0.7}Co_{0.3}O_2$, $LiFePO_4$, $LiMnPO_4$ and $LiNi_{0.5}Mn_{1.5}O_4$ cathodes; showing better in the reversible capacity of the composite electrodes compared to other carbon polymorphs [62–65].

Lithium Sulfur Batteries (Li-SBs): After LIBs, Lithium sulfur batteries are drawing much attention due to the high energy density of lithium-sulfur (Li-S) batteries (2600 Wh kg^{-1}) and is natural abundance of sulfur. Beside the potential advantages, the major challenge is the electronically insulation behavior of sulfur. In addition, during the cycling processes, the polysulfides are formed which are soluble, and discharge intermediate and products migrate towards Li anode. This impacts the columbic efficiency, accelerates battery self-discharge and cycle life. Several research groups are using CNTs for sulfur encapsulation to overcome above mentioned problems [66]. The sulfur is incorporated carbon nanotubes, nano pores and/or in between nano wires for Li-S battery cathode. The electrode delivers discharge capacity of 669 mAh g^{-1} after 300 cycles with a low capacity fading rate of 0.166% per cycle at 0.1 C rate [67]. Sometimes functional groups were grafted on the modified multi-walled carbon nanotubes which can adsorb the dissolved polysulfides and enhance the redox reaction of lithium polysulfides and in parallel provides the electronic conduction pathway.

Sodium Ion Batteries (SIBs): Off significance, CNTs cannot be used as anode for Na ion batteries, like LIBs, because of large radius of Na ion (1.02 Å) which cannot be intercalated comfortably into the layer structure. The Na ion intercalation into graphite is thermodynamically unstable and it cannot form primary stage structures of NaC_6 or NaC_8. The Pure graphite can deliver a maximum capacity of ~31 mAh g^{-1} by forming NaC_{70} [68]. The defect-rich and disordered carbon nanotube structures have been synthesized for enhance the sodium storage as an anode for SIBs, which exhibits reversible capacities over 130 mAh g^{-1}. Very recently, Han et al., prepared high defective and disorder mesoporous carbon nanotubes by ethanol flame method. The electrode displays a remarkable rate capability of 145 mAh g^{-1} at 1 A g^{-1}, with excellent cyclability [69]. Another approach to obtain defective CNTs is doping of heteroatoms, such as nitrogen, which can also enhance the electrical conductivity of carbon nanotubes [70].

CNTs have been using as an additive for lower electronic conductive electrode materials in SIBs. It was reported that porous $FePO_4$ nanoparticles were electrically connected by single-wall carbon nanotubes synthesized by hydrothermal reaction. The fabricated composite electrode shows discharge capacity of 120 mAh g^{-1} at a 0.1 C rate with unprecedented cycling stability [71]. The CNTs have been using as a promising additive for polyhedral cathode materials like $NaTi_2(PO_4)_3$, $Na_3V_2(PO_4)_2F_3$, Na_2FePO_4F, $NaVPO_4F$, $Na_4VMn(PO_4)_3$, $Na_4MnCr(PO_4)_3$, $Na_3V_2(PO_4)_3$, $Na_2Fe(SO_4)_2$, Na_2MnSiO_4, $Na_3V_2O_{2x}(PO_4)_2F_{3-2x}$, $Na_4Co_3(PO_4)_2P_2O_7$, Prussian blue analogues ...etc. [72]. Our group published the impact of MWCNT on particle growth as well as electrochemical properties of $Na_3V_2O_{2x}(PO_4)_2F_{3-2x}$ cathode. Among three carbon sources (Carbon, MWCNT & rGO), MWCNT is more effective to obtain moderate particle size with enhanced electrochemical properties (**Figure 5**). The prepared $Na_3V_2O_{2x}(PO_4)_2F_{3-2x}$-MWCNT composite delivers the stable capacity of 98 and 89 mAh g^{-1} in half cell and full cell with $NaTi_2(PO_4)_3$-MWCNT configurations, respectively [73]. It should be noted that most of the alloying and conversion anode materials lose their electron conducting path due to the pulverization during charge–discharge cycles. In this case, CNT can be used as conductive additive as well as electrode integrity protector. The battery research community has been encapsulated metal based (e.g. Sn) anode with the CNTs to accommodate the volume expansion during Na insertion to avoid the pulverization. The reported results indicate that the carbon encapsulated, Sn@N-doped, nanotubes is beneficial to get good reversible capacity of 398 mAh g^{-1} at 100 mA g^{-1}, with capacity retention of 67% over 150 cycles [74, 75]. The ultrathin MoS_2 nanosheets was developed on the surfaces of CNTs by a hydrothermal method MoS_2/CNTs, which exhibit excellent electrochemical performance as conversion anode materials for SIBs. The MoS_2/CNTs, shows a reversible capacity of 504 mAh g^{-1} at a current rate of 50 mA g^{-1} over 100 cycles [76]. Many alloying and conversion anode materials have used CNTs as conductive additive, examples TiO_2, MoS_2, CuS, Fe_2O_3, & FeO.

Potassium Ion Batteries (PIBs): Unlike sodium, potassium ion can be intercalated into graphite structure without requiring a special electrolyte solvent for K-ion batteries (PIBs). It was reported that theoretical capacity of K battery is 279 mAh g^{-1} (KC_8) by stepwise potassiation through KC_{36} and KC_{24} phases based on

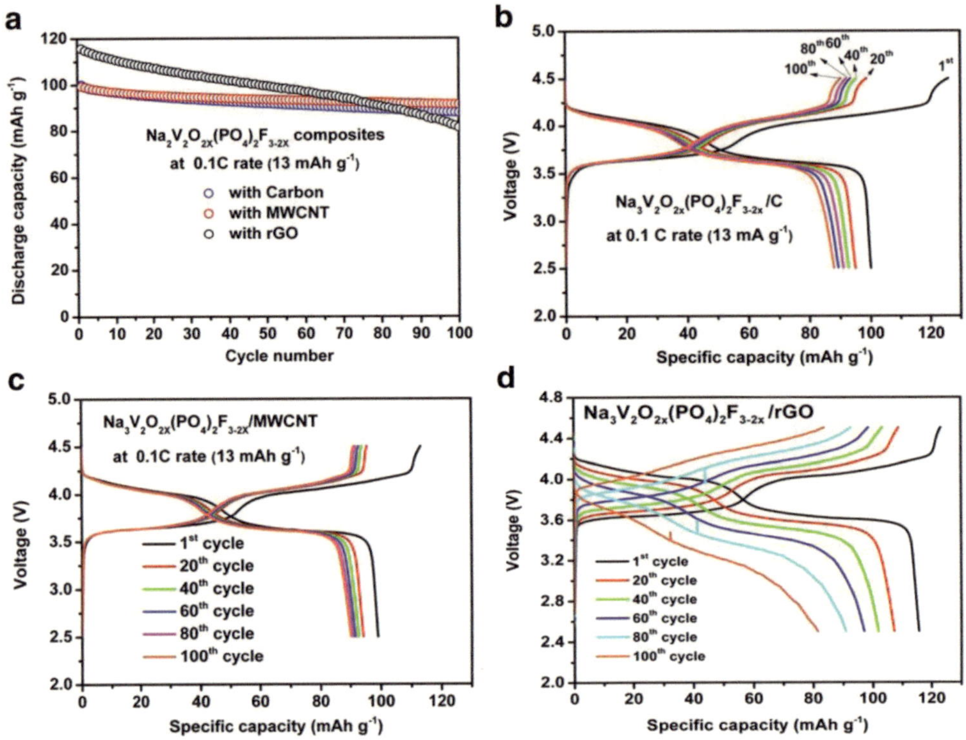

Figure 5.
(a) Cyclability of Na3V2O2x(PO4)2F3-2x along with three different carbon materials. Charge-discharge curves for the Na3V2O2x(PO4)2F3-2x with (b) carbon, (c) MWCNT, and (d) rGO. Reproduced from Ref. [73] with permission from Springer Nature.

intercalation/deintercalation mechanism [77]. Noting that the insertion potential of K^+ into the graphite structure is little higher than that of Li^+, which could make more secure battery systems. However, the biggest obstacle is the poor cycle stability of graphite as the anode for PIBs. Battery community have been trying to improve the performances of PIBs and fulfill the requirements of commercialization [78]. Liu *et al.*, prepared N-doped bamboo-like carbon nanotubes by simple pyrolysis method and the unique structured material shows a high reversible capacities of 204 mAh g^{-1} and 186 mAh g^{-1} at 500 mA g^{-1} and 1000 mA g^{-1}, respectively [79]. The science behind the better performance is not well understood yet.

The analysis of electron density difference demonstrates the interaction between the K ion and the nitrogen doped CNTs which has strong ionic bonding, and the electron re-distributions between N5 & N6 CNTs. It is shown, in the K ion –N5 CNT systems (**Figure 6A**), the net gain of electronic charge on the pyrrolic N atom plays more significant role than those of the other two pyridinic N atoms. The N6 CNT (**Figure 6B**), the alkali metal atom associates strongly with two pyridinic N atoms, therefore, the overlapping of the corresponding peaks in **Figure 6** (bottom) is seen. The bonding with the third pyridinic N atom is relatively weaker [80]. The theoretical studies predicted that inner carbon of CNT is dense while outer carbon of CNT is loosely bind. The hierarchical carbon nanotubes structures in the inner dense part act as skeleton while the outer loose-CNT effectively accommodates the K-ion accommodation, which are showing a better specific capacity of 232 mAh g^{-1} and good cyclic stability [81]. Like other electrode systems these carbon nanotubes are expected to act as a conducting additive assuring the electrical percolation in the composite electrode and to protect the integrity of electrode using their mechanical properties [82, 83].

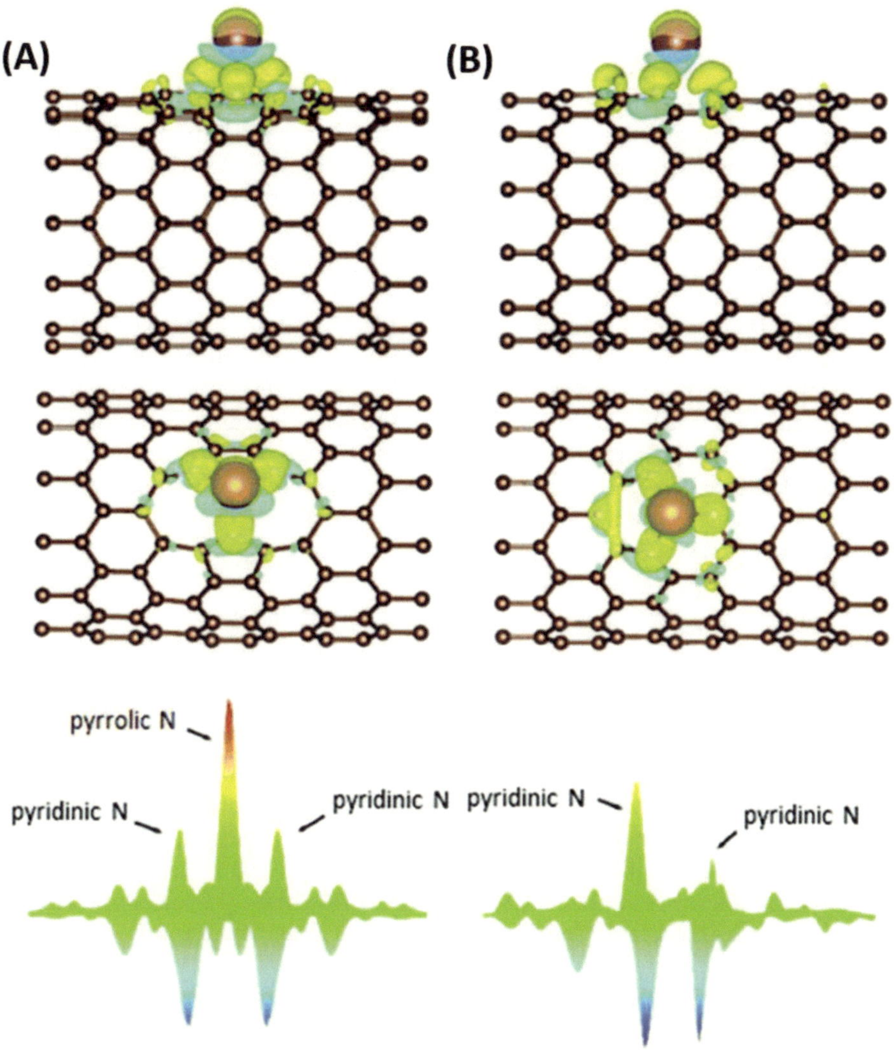

Figure 6.
Differential electron densities (A) K-ion on N5 CNT, (B) K-ion on N6 CNT: top, side view; middle, top view; bottom, electron density differences in the plane. Reproduced from Ref. [80] with permission from the Royal Society of Chemistry.

Lead acid batteries*:* Lead acid battery is one of the most popular electrochemical storage systems for the last 150-years, however, it has been suffering from poor lifecycle. The limited lifecycle is most probably originated due to the formation of large non-conducting uncontrolled lead sulfate ($PbSO_4$) crystals both the positive and negative plates. The deposition of insulating $PbSO_4$ crystals lower the electrical conductivity and accessibility of electrolytes to active material in both plates [84, 85]. Various research groups studies different amorphous carbons as a sulfation-suppressing additive in negative plates, because the sulfation is more prominent in negative plate than positive plate due to slower kinetic reaction. Recently, Prof. Aurbach and his group used SWCNTs as a suppresser of uncontrolled sulfation processes in lead-acid battery electrodes. The carbon nanotubes additive would be uniformly distributed throughout the composite electrode and capable of boosting charge acceptance at low concentrations [86, 87].

4.2 Metal-air batteries

Lithium-Air Batteries (LABs*):* Recently battery community focused on the metal-air battery due to higher theoretical density. It is just an alternative to LIBs. The most popular and promising metal–air batteries are lithium -air and zinc-air batteries. The energy density of rechargeable lithium-air batteries very high energy (~1700 Wh kg^{-1}) comparable to gasoline and much higher than secondary Li-ion batteries (~160 Wh kg^{-1}). The reaction mechanism of lithium air battery is appeared to be very simple, at discharge state oxygen in air reduced by lithium ions to form lithium peroxides via $2Li^+ + 2e^- + O_2 \Leftrightarrow Li_2O_2$ and/or $4Li^+ + 4e^- + O_2 \Leftrightarrow 2Li_2O$, and formation of lithium and oxygen from decomposition of lithium oxides during charge processes. The thermodynamic equilibrium cell voltage for the discharge reaction in LABs is 2.96 V vs. Li/Li^+ [88]. In practical realization, the reported cell voltage is less than 2.96 V which is due to the cell polarization resulted from the oxygen reduction and evolution reaction during discharge and charge processes. However, breaking the discharge products during the charge processes, it requires much more than 2.96 V to drive the reverse electrochemical reaction. Either pure catalyst or carbon -supported catalyst particles have been used to accelerate the electrode reactions [89]. It should be mentioned that most of the time CNTs have been used as conductive supporting materials for metal and metal oxide catalyst particles in the metal-air batteries. The functionalized CNTs can also be used as air electrodes. It was reported that the flexible multiwalled carbon nanotube exhibited very high specific capacity of 34,600 mAh g^{-1} at a current density of 500 mA g^{-1} in the $Li–O_2$ batteries [90]. It is indicated that CNTs have huge prospectuses as in the Li-air battery cathode component.

Zn-Air batteries (ZAB): Zinc–air batteries are very safe for electrical vehicles which is fabricated by non-flammable and non-explosive materials. They can be used for other safe applications. As mentioned above, the electrocatalysts is required in air electrode to efficiently accelerate the kinetics of the oxygen reactions [91] and increases the battery performances and efficiencies. It is demonstrated that the nitrogen-doped carbon nanotubes (N-CNTs) promoted notable ORR activity in acid and alkaline solutions. This is because of the inserted heterogeneous nitrogen which might activates the reaction sites and can induce in breaking the O-O bonds of O_2 molecules [92].

Another critical role of CNTs in batteries is the current collector. Present, flexible CNTs based carbon papers can be fabricated from all CNTs and used as anode and current collector for aqueous battery systems. Conventional current collectors, such as carbon cloth and metal foils (stainless steel, Titanium), are low surface area and highly corrosive in aqueous media. Also, these CNTs can be used as a pure binder in primary thermal battery electrode fabrication. The electrode with the CNTs binder has better thermal stability than conventional organic binders. The traditional organic binders were decomposed before reaching the operating temperature of 500°C, and its residual material can act as an insulator.

5. Fuel-cells (FCs)

The reaction mechanism in Li-air battery and fuel cells has great similarity where oxygen reduction reaction (ORR) and oxygen evolution reaction (OER) are important for fuel cells efficiency. To enhance the efficiency of the fuel cell, a catalyst is needed. Instead of using expensive Pt as a catalyst, researchers started using a supporter, which can improve the capability of low-cost catalyst. Commonly used catalysts supporters are porous carbon, carbon nanotubes, graphene, and other carbon polymorphs. It was demonstrated, at higher current density, CNT supported FCs, exhibited better electrochemical performances than the carbon black supported FCs [93]. Doping with heteroatoms or loading of transition metal catalysts on CNTs substantially enhance the activity of highly efficient fuel cells. There are few reports on encapsulation of Ag, Fe, Co, CuSe (**Figure 7**) & Ni based compounds in pure CNTs, which are showing the high ORR performance in fuel cells [94]. It is also reported that the higher oxidation state of Ni is very active for OER and inactive to ORR. However, Ni encapsulated N doped CNTs are showing very high ORR activity and less OER active. Several studies are compared the performances of the platinum catalyst with non-noble metal catalysts with the CNT support and they exhibit better catalytic activity and it reduces the cost of whole cell. Furthermore, CNTs can make the fuel cell highly stable and high resistive against corrosion during electrochemical reaction [94, 95]. CNTs not only increase the catalytic activity; enhance the corrosion resistance. Besides, CNTs improve the mass transmission capability of both electrodes in a fuel cell.

5.1 Supercapacitors

The morphology of electrode materials and fabrication process plays an important role for the performance of a supercapacitor. The capacitance value of a supercapacitor is highly dependent on electrode surface-area and porosity. The basic principle of a capacitor is to store energy by separation of charge at the electrode and electrolyte interface (i.e., double layer capacitance). The ions transfer between the two electrodes is mediated by diffusion across the electrolyte [96]. Supercapacitors exhibits better reversibility, higher power density, and longer cycle life which made it attentive and promising for energy-storage devices. It is worth to mention that supercapacitors exhibit the highest known power capability (2–5 kW kg^{-1}), but they suffer from a moderate energy density (3–6 Wh kg^{-1}).

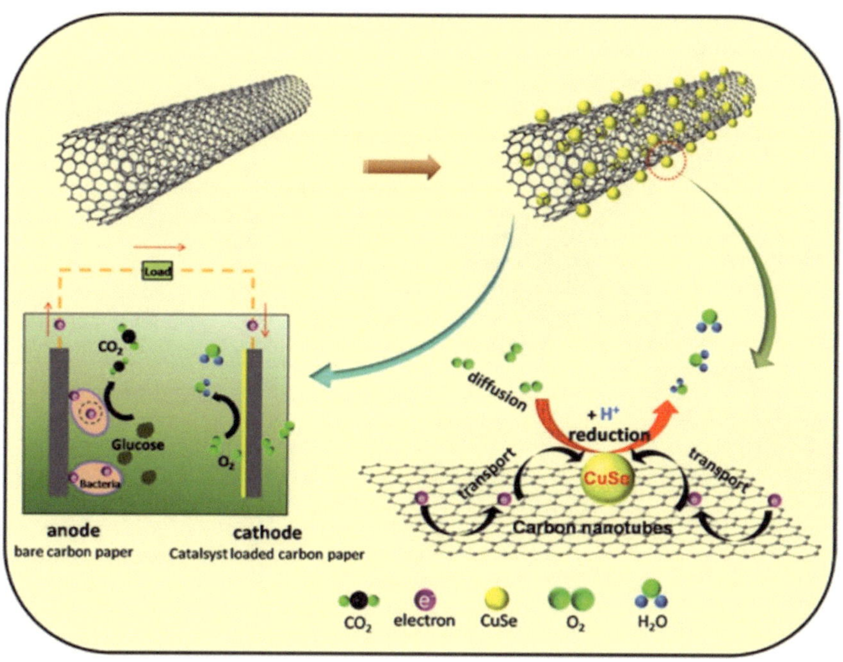

Figure 7.
Carbon nanotubes decorated with copper selenide (CuSe) nanoparticles for microbial fuel cells. Reproduced from Ref. [94] with permission from Elsevier.

Carbon nanotubes (CNTs) are very promising as supercapacitor electrode materials because of their excellent electrical properties and one-dimensional nanostructures. Noting that defect free or less defect CNTs has smaller surface area and micropore content than conventional activated carbon (AC), which made them insufficient capacitance in CNT-based electrodes. However, it is reported that the formation of defects on surface and open ends by alkaline solution activation increases the surface area of CNTS [97] and exhibits better capacitance value. The SWCNTs show enhanced specific capacitance than those of MWCNTs which results from large surface area of SWCNTs. However, that MWCNTs could generate capacitance twice as high in comparison to SWCNTs which is attributed to the presence of mesopores and entangled tube structure, facilitating the transport of the ions [98]. The flexible aligned SWCNTs with high surface area and better electrical conductivity is beneficial for capacitors applications [99]. It should be mentioned that contact resistance reduces the performance of supercapacitor and therefore, polished metal foils is used as current collectors to grow the carbon nanotubes for lowering contact resistance. The better discharge efficiency can be obtained through the electrodynamics and can result high power density [100]. The cell resistance can be lower either by fabricating carbon nanotubes as thin film electrodes which has coherent structures with highly concentrated colloidal suspension or fabricating CNT based thin film electrodes using an electrophoretic deposition (EPD) method. It is reported that these flexible CNTs films are binder free and forms network with negligible electrode resistance [101]. As we mentioned in above applications, N doped CNTs may contribute to improving the power characteristics of supercapacitors their own way. The doped nitrogen modifies the conduction band and the modified electronic structure which helps to enhance the quantum capacitance and electrical conductivity of CNTs [102]. Recently researchers have started the fabrication of a high-performance wire-type supercapacitor with CNTs to get the high voltage and high energy density (**Figure 8**). It should be noted that the carbon nanotube sheets were wrapped to make a fiber shaped supercapacitors on elastic polymeric fibers with moderate stretch ability [103, 104].

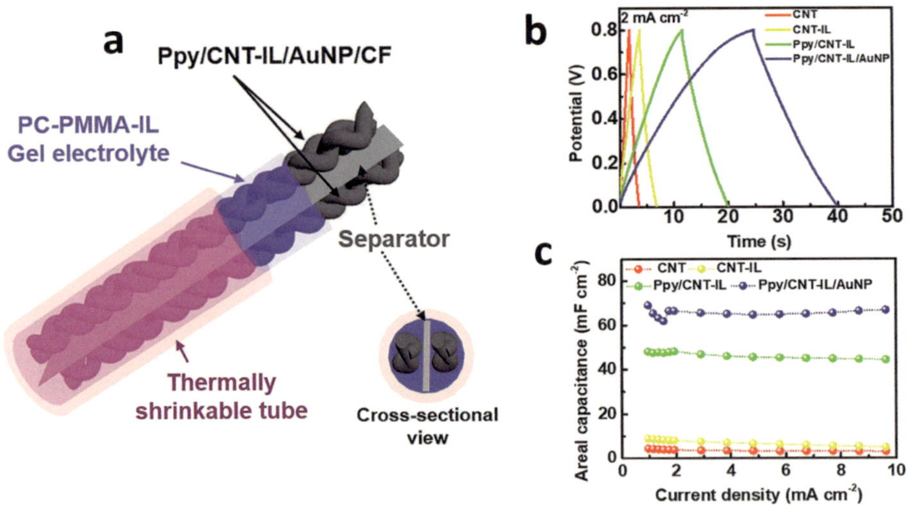

Figure 8.
(a) Schematic representation of the wire-type supercapacitor, (b) galvanostatic charge/discharge curves and (c) Comparison plots of areal capacitance versus current density for CF electrodes coated with CNT, CNT-IL, Ppy/CNT-Ionic Liquid, and Ppy/CNT-Ionic Liquid/AuNP. Reproduced from Ref. [104] with permission from Elsevier.

Graphitization and pore size distribution of CNT are also significant factors for supercapacitor application. While heating, the specific surface area increases, but the capacitance decreased due to the average pore diameter decreases and saturated at high temperature. Furthermore, chemically activated of CNTs also shows tubular morphology with defects on the surface that gave a significant increase in pore volume. Aligned CNTs can also significantly improve the capacitance and power density of supercapacitors. It is also reported that the highly packed and aligned CNTs showed higher capacitance and less capacitance drop when compared to other thick CNT based electrodes.

6. Conclusions

One-dimensional carbon nanotubes (CNTs) have been considered as potential candidates for the development of energy storage materials based on their unique chemical and physical properties. The architecture and quality of the CNTs plays a vital role on the electrochemical performances exhibited by both batteries and supercapacitors. It is observed that a slight modification (defects creation, hetero-atoms doping & controlling the distribution of pore sizes) in the CNT structure brings out complementary properties that translate to excellent electrochemical performances. Anchored and directly grown aligned structure of CNTs trends to have high stability and fast ion transportation. The composite electrode with incorporated CNTs is being benefited from the high surface area, excellent conductivity, enhanced specific capacity, better cyclability and rate capability. CNTs can be used as an electrochemically active and inactive electrode component in energy storage systems. It turns out that all types of CNTs can serve as flexible supporting materials and can also enable next generation flexible energy storage devices. The future of advanced energy storage systems (either batteries or supercapacitor) can certainly be benefited from the incorporation of CNTs. The extraordinarily high electronic conductivity also enables CNTs and graphite as an additive to the composite electrode and enable to activate poorly conducting electrode materials to make them electrochemically active. Moreover, the structures and morphologies of CNTs are beneficial for supercapacitors and as catalyst support for fuel cells.

Acknowledgements

This manuscript has been supported by Oak Ridge Nation Laboratory (ORNL) managed by UT-Battelle, LLC, under contract DE-AC05-00OR22725 with the US Department of Energy (DOE).

Author details

Ruhul Amin[1]*, Petla Ramesh Kumar[2] and Ilias Belharouak[1]

1 Energy and Transportation Science Division, Oak Ridge National Laboratory, Oak Ridge, TN, USA

2 Department of Applied Chemistry, Tokyo University of Science, Tokyo, Japan

*Address all correspondence to: aminr@ornl.gov

References

[1] Iijima S, Ichihashi T, Ando Y. Pentagons, heptagons and negative curvature in graphite microtubule growth. Nature. 1992;356(6372):776-778.

[2] Iijima S. Helical microtubules of graphitic carbon. Nature. 1991;354(6348):56-58.

[3] Iijima S, Ajayan PM, Ichihashi T. Growth model for carbon nanotubes. Phys Rev Lett. 1992;69(21):3100-3103.

[4] Iijima S, Ichihashi T. Single-shell carbon nanotubes of 1-nm diameter. Nature. 1993;363(6430):603-605.

[5] Yu M-F, Lourie O, Dyer MJ, Moloni K, Kelly TF, Ruoff RS. Strength and Breaking Mechanism of Multiwalled Carbon Nanotubes Under Tensile Load. Science. 2000;287(5453):637.

[6] Ando Y, Zhao X, Shimoyama H, Sakai G, Kaneto K. Physical properties of multiwalled carbon nanotubes. Int J Inorg Mater. 1999;1(1):77-82.

[7] Zhang Y, Bunes BR, Wu N, Ansari A, Rajabali S, Zang L. Sensing methamphetamine with chemiresistive sensors based on polythiophene-blended single-walled carbon nanotubes. Sensors and Actuators B: Chemical. 2018;255:1814-1818.

[8] Shobin LR, Manivannan S. Silver nanowires-single walled carbon nanotubes heterostructure chemiresistors. Sensors and Actuators B: Chemical. 2018;256:7-17.

[9] Rakhi RB, Sethupathi K, Ramaprabhu S. Electron field emission properties of conducting polymer coated multi walled carbon nanotubes. Appl Surf Sci. 2008;254(21):6770-6774.

[10] Yan Y, Miao J, Yang Z, Xiao F-X, Yang HB, Liu B, et al. Carbon nanotube catalysts: recent advances in synthesis, characterization and applications. Chem Soc Rev. 2015;44(10):3295-3346.

[11] Saito S. Carbon Nanotubes for Next-Generation Electronics Devices. Science. 1997;278(5335):77.

[12] Landi BJ, Ganter MJ, Cress CD, DiLeo RA, Raffaelle RP. Carbon nanotubes for lithium ion batteries. Energy Environ Sci. 2009;2(6):638-654.

[13] Akbari E, Buntat Z. Benefits of using carbon nanotubes in fuel cells: a review. International Journal of Energy Research. 2017;41(1):92-102.

[14] Collins PG, Zettl A, Bando H, Thess A, Smalley RE. Nanotube Nanodevice. Science. 1997;278(5335):100.

[15] Pan H, Li J, Feng Y. Carbon nanotubes for supercapacitor. Nanoscale research letters. 2010;5(3):654-668.

[16] Cheng H-M, Yang Q-H, Liu C. Hydrogen storage in carbon nanotubes. Carbon. 2001;39(10):1447-1454.

[17] Dresselhaus MS, Dresselhaus G, Saito R. Physics of carbon nanotubes. Carbon. 1995;33(7):883-891.

[18] Monea BF, Ionete EI, Spiridon SI, Ion-Ebrasu D, Petre E. Carbon Nanotubes and Carbon Nanotube Structures Used for Temperature Measurement. Sensors. 2019;19(11).

[19] Amelinckx S, Lucas A, Lambin P. Electron diffraction and microscopy of nanotubes. Rep Prog Phys. 1999;62(11):1471-1524.

[20] Belin T, Epron F. Characterization methods of carbon nanotubes: a review. Materials Science and Engineering: B. 2005;119(2):105-118.

[21] Hodge SA, Bayazit MK, Coleman KS, Shaffer MSP. Unweaving the rainbow: a review of the relationship between single-walled carbon nanotube molecular structures and their chemical reactivity. Chem Soc Rev. 2012;41(12):4409-4429.

[22] Zandiatashbar A, Lee G-H, An SJ, Lee S, Mathew N, Terrones M, et al. Effect of defects on the intrinsic strength and stiffness of graphene. Nature Comm. 2014;5(1):3186.

[23] Terrones H, Lv R, Terrones M, Dresselhaus MS. The role of defects and doping in 2D graphene sheets and 1D nanoribbons. Rep Prog Phys. 2012;75(6):062501.

[24] Wei BQ, Vajtai R, Ajayan PM. Reliability and current carrying capacity of carbon nanotubes. Appl Phys Lett. 2001;79(8):1172-1174.

[25] Pugno NM. The role of defects in the design of space elevator cable: From nanotube to megatube. Acta Mater. 2007;55(15):5269-5279.

[26] Robertson DH, Brenner DW, Mintmire JW. Energetics of nanoscale graphitic tubules. Physical Review B. 1992;45(21):12592-12595.

[27] Hone J, Batlogg B, Benes Z, Johnson AT, Fischer JE. Quantized Phonon Spectrum of Single-Wall Carbon Nanotubes. Science. 2000;289(5485):1730.

[28] Yamashita S. Nonlinear optics in carbon nanotube, graphene, and related 2D materials. APL Photonics. 2018;4(3):034301.

[29] Matsuda K. 1 - Fundamental optical properties of carbon nanotubes and graphene. In: Yamashita S, Saito Y, Choi JH, editors. Carbon Nanotubes and Graphene for Photonic Applications: Woodhead Publishing; 2013. p. 3-25.

[30] Eder D. Carbon Nanotube–Inorganic Hybrids. Chem Rev. 2010;110(3):1348-1385.

[31] Xiong Z, Yun YS, Jin H-J. Applications of Carbon Nanotubes for Lithium Ion Battery Anodes. Materials. 2013;6(3).

[32] de las Casas C, Li W. A review of application of carbon nanotubes for lithium ion battery anode material. J Power Sources. 2012;208:74-85.

[33] Shimoda H, Gao B, Tang XP, Kleinhammes A, Fleming L, Wu Y, et al. Lithium Intercalation into Opened Single-Wall Carbon Nanotubes: Storage Capacity and Electronic Properties. Phys Rev Lett. 2001;88(1):015502.

[34] Nishidate K, Hasegawa M. Energetics of lithium ion adsorption on defective carbon nanotubes. Physical Review B. 2005;71(24):245418.

[35] Mi CH, Cao GS, Zhao XB. A non-GIC mechanism of lithium storage in chemical etched MWNTs. Journal of Electroanalytical Chemistry. 2004;562(2):217-221.

[36] Yang S, Huo J, Song H, Chen X. A comparative study of electrochemical properties of two kinds of carbon nanotubes as anode materials for lithium ion batteries. Electrochim Acta. 2008;53(5):2238-2244.

[37] Meunier V, Kephart J, Roland C, Bernholc J. Ab Initio Investigations of Lithium Diffusion in Carbon Nanotube Systems. Phys Rev Lett. 2002;88(7):075506.

[38] Prem Kumar T, Ramesh R, Lin YY, Fey GT-K. Tin-filled carbon nanotubes as insertion anode materials for lithium-ion batteries. Electrochem Commun. 2004;6(6):520-525.

[39] Wang XX, Wang JN, Su LF. Preparation and electrochemical

performance of ultra-short carbon nanotubes. J Power Sources. 2009;186(1):194-200.

[40] Wang XX, Wang JN, Chang H, Zhang YF. Preparation of Short Carbon Nanotubes and Application as an Electrode Material in Li-Ion Batteries. Adv Funct Mater. 2007;17(17):3613-3618.

[41] Garau C, Frontera A, Quiñonero D, Costa A, Ballester P, Deyà PM. Ab initio investigations of lithium diffusion in single-walled carbon nanotubes. Chem Phys. 2004;297(1):85-91.

[42] Zhao M, Xia Y, Liu X, Tan Z, Huang B, Li F, et al. Curvature-induced condensation of lithium confined inside single-walled carbon nanotubes: First-principles calculations. Phys Lett A. 2005;340(5):434-439.

[43] Hosam MS, Martin K. Introductory Chapter: Carbon Nanotubes. 2019.

[44] Carter R, Oakes L, Cohn AP, Holzgrafe J, Zarick HF, Chatterjee S, et al. Solution Assembled Single-Walled Carbon Nanotube Foams: Superior Performance in Supercapacitors, Lithium-Ion, and Lithium–Air Batteries. J Phys Chem, C. 2014;118(35):20137-20151.

[45] Eom J-Y, Kwon H-S. Effects of the chemical etching of single-walled carbon nanotubes on their lithium storage properties. Mater Chem Phys. 2011;126(1):108-113.

[46] Eom J, Kim D, Kwon H. Effects of ball-milling on lithium insertion into multi-walled carbon nanotubes synthesized by thermal chemical vapour deposition. J Power Sources. 2006;157(1):507-514.

[47] DiLeo RA, Castiglia A, Ganter MJ, Rogers RE, Cress CD, Raffaelle RP, et al. Enhanced Capacity and Rate Capability of Carbon Nanotube Based Anodes with Titanium Contacts for Lithium Ion Batteries. ACS Nano. 2010;4(10):6121-6131.

[48] Liu X-M, Huang Zd, Oh Sw, Zhang B, Ma P-C, Yuen MMF, et al. Carbon nanotube (CNT)-based composites as electrode material for rechargeable Li-ion batteries: A review. Compos Sci Technol. 2012;72(2):121-144.

[49] Zhao J, Buldum A, Han J, Ping Lu J. First-Principles Study of Li-Intercalated Carbon Nanotube Ropes. Phys Rev Lett. 2000;85(8):1706-1709.

[50] Zhang H, Cao G, Wang Z, Yang Y, Shi Z, Gu Z. Carbon nanotube array anodes for high-rate Li-ion batteries. Electrochim Acta. 2010;55(8):2873-2877.

[51] Chew SY, Ng SH, Wang J, Novák P, Krumeich F, Chou SL, et al. Flexible free-standing carbon nanotube films for model lithium-ion batteries. Carbon. 2009;47(13):2976-2983.

[52] Lahiri I, Oh S-W, Hwang JY, Cho S, Sun Y-K, Banerjee R, et al. High Capacity and Excellent Stability of Lithium Ion Battery Anode Using Interface-Controlled Binder-Free Multiwall Carbon Nanotubes Grown on Copper. ACS Nano. 2010;4(6):3440-3446.

[53] Masarapu C, Subramanian V, Zhu H, Wei B. Long-Cycle Electrochemical Behavior of Multiwall Carbon Nanotubes Synthesized on Stainless Steel in Li Ion Batteries. Adv Funct Mater. 2009;19(7):1008-1014.

[54] Li J, Kaur AP, Meier MS, Cheng Y-T. Stacked-cup-type MWCNTs as highly stable lithium-ion battery anodes. Journal of Applied Electrochemistry. 2014;44(1):179-187.

[55] Landi BJ, Ganter MJ, Schauerman CM, Cress CD, Raffaelle RP. Lithium Ion Capacity of Single Wall Carbon Nanotube

Paper Electrodes. J Phys Chem, C. 2008;112(19):7509-7515.

[56] Wang L, Guo W, Lu P, Zhang T, Hou F, Liang J. A Flexible and Boron-Doped Carbon Nanotube Film for High-Performance Li Storage. Frontiers in Chemistry. 2019;7(832).

[57] Li X, Liu J, Zhang Y, Li Y, Liu H, Meng X, et al. High concentration nitrogen doped carbon nanotube anodes with superior Li+ storage performance for lithium rechargeable battery application. J Power Sources. 2012;197:238-245.

[58] Wang B, Li X, Luo B, Zhang X, Shang Y, Cao A, et al. Intertwined Network of Si/C Nanocables and Carbon Nanotubes as Lithium-Ion Battery Anodes. ACS Appl Mater Interfaces. 2013;5(14):6467-6472.

[59] Zhang H, Wang Y, Zhao W, Zou M, Chen Y, Yang L, et al. MOF-Derived ZnO Nanoparticles Covered by N-Doped Carbon Layers and Hybridized on Carbon Nanotubes for Lithium-Ion Battery Anodes. ACS Appl Mater Interfaces. 2017;9(43):37813-37822.

[60] Bhandavat R, Singh G. Stable and Efficient Li-Ion Battery Anodes Prepared from Polymer-Derived Silicon Oxycarbide–Carbon Nanotube Shell/Core Composites. J Phys Chem, C. 2013;117(23):11899-11905.

[61] Landi BJ, Raffaelle RP, Heben MJ, Alleman JL, VanDerveer W, Gennett T. Single Wall Carbon Nanotube−Nafion Composite Actuators. Nano Lett. 2002;2(11):1329-1332.

[62] Jin EM, Jin B, Park KH, Gu H-B, Park G-C, Kim K-W. Electrochemical Characteristics of Lithium Iron Phosphate with Multi-Walled Carbon Nanotube for Lithium Polymer Batteries. J Nanosci Nanotech. 2008;8:5057-5061.

[63] Guoping W, Qingtang Z, Zuolong Y, MeiZheng Q. The effect of different kinds of nano-carbon conductive additives in lithium ion batteries on the resistance and electrochemical behavior of the LiCoO2 composite cathodes. Solid State Ionics. 2008;179(7):263-268.

[64] Chiu T-M, Barraza-Fierro JI, Castaneda H. Comprehensive Interfacial Mechanisms of LiMnPO4-MWCNT Composite ratios in Acidic Aqueous Electrolyte. Electrochim Acta. 2017;253:93-103.

[65] Van Le T, Le MLP, Van Tran M, Nguyen NMT, Luu AT, Nguyen HT. Fabrication of Cathode Materials Based on Limn2o4/Cnt and Lini0.5mn1.5o4/Cnt Nanocomposites for Lithium Ion Batteries Application. Materials Research. 2015;18:1044-1052.

[66] Shi Z, Yang Y, Huang Y, Yue H, Cao Z, Dong H, et al. Organic Alkali Metal Salt Derived Three-Dimensional N-Doped Porous Carbon/Carbon Nanotubes Composites with Superior Li–S Battery Performance. ACS Sustainable Chem Eng. 2019;7(4):3995-4003.

[67] Xu Y-W, Zhang B-H, Li G-R, Liu S, Gao X-P. Covalently Bonded Sulfur Anchored with Thiol-Modified Carbon Nanotube as a Cathode Material for Lithium–Sulfur Batteries. ACS Applied Energy Materials. 2020;3(1):487-494.

[68] Xiao B, Rojo T, Li X. Hard Carbon as Sodium-Ion Battery Anodes: Progress and Challenges. ChemSusChem. 2019;12(1):133-144.

[69] Han W, Zhou Y, Zhu T, Chu H. Combustion synthesis of defect-rich carbon nanotubes as anodes for sodium-ion batteries. Appl Surf Sci. 2020;520:146317.

[70] Li D, Zhang L, Chen H, Ding L-x, Wang S, Wang H. Nitrogen-doped

bamboo-like carbon nanotubes: promising anode materials for sodium-ion batteries. Chem Commun. 2015;51(89):16045-16048.

[71] Liu Y, Xu Y, Han X, Pellegrinelli C, Zhu Y, Zhu H, et al. Porous Amorphous FePO4 Nanoparticles Connected by Single-Wall Carbon Nanotubes for Sodium Ion Battery Cathodes. Nano Lett. 2012;12(11):5664-5668.

[72] Kumar PR, Essehli R, Yahia HB, Amin R, Belharouak I. Electrochemical studies of a high voltage Na4Co3(PO4)2P2O7–MWCNT composite through a selected stable electrolyte. RSC Adv. 2020;10(27):15983-15989.

[73] Kumar PR, Jung YH, Kim DK. Influence of carbon polymorphism towards improved sodium storage properties of Na3V2O2x(PO4)2F3-2x. J Solid State Electrochem. 2017;21(1):223-232.

[74] Luu THT, Duong DL, Lee TH, Pham DT, Sahoo R, Han G, et al. Monodispersed SnS nanoparticles anchored on carbon nanotubes for high-retention sodium-ion batteries. J Mater Chem A. 2020;8(16):7861-7869.

[75] Ruan B, Guo H-p, Hou Y, Liu Q, Deng Y, Chen G, et al. Carbon-Encapsulated Sn@N-Doped Carbon Nanotubes as Anode Materials for Application in SIBs. ACS Appl Mater Interfaces. 2017;9(43):37682-37693.

[76] Zhang S, Yu X, Yu H, Chen Y, Gao P, Li C, et al. Growth of Ultrathin MoS2 Nanosheets with Expanded Spacing of (002) Plane on Carbon Nanotubes for High-Performance Sodium-Ion Battery Anodes. ACS Appl Mater Interfaces. 2014;6(24):21880-21885.

[77] Jian Z, Luo W, Ji X. Carbon Electrodes for K-Ion Batteries. J Am Chem Soc. 2015;137(36):11566-11569.

[78] Fan L, Ma R, Zhang Q, Jia X, Lu B. Graphite Anode for a Potassium-Ion Battery with Unprecedented Performance. Angew Chem Int Ed. 2019;58(31):10500-10505.

[79] Liu Y, Yang C, Pan Q, Li Y, Wang G, Ou X, et al. Nitrogen-doped bamboo-like carbon nanotubes as anode material for high performance potassium ion batteries. J Mater Chem A. 2018;6(31):15162-15169.

[80] Zhao C, Lu Y, Liu H, Chen L. First-principles computational investigation of nitrogen-doped carbon nanotubes as anode materials for lithium-ion and potassium-ion batteries. RSC Adv. 2019;9(30):17299-17307.

[81] Wang Y, Wang Z, Chen Y, Zhang H, Yousaf M, Wu H, et al. Hyperporous Sponge Interconnected by Hierarchical Carbon Nanotubes as a High-Performance Potassium-Ion Battery Anode. Adv Mater. 2018;30(32):1802074.

[82] Zeng S, Zhou X, Wang B, Feng Y, Xu R, Zhang H, et al. Freestanding CNT-modified graphitic carbon foam as a flexible anode for potassium ion batteries. J Mater Chem A. 2019;7(26):15774-15781.

[83] Gabaudan V, Touja J, Cot D, Flahaut E, Stievano L, Monconduit L. Double-walled carbon nanotubes, a performing additive to enhance capacity retention of antimony anode in potassium-ion batteries. Electrochem Commun. 2019;105:106493.

[84] Marom R, Ziv B, Banerjee A, Cahana B, Luski S, Aurbach D. Enhanced performance of starter lighting ignition type lead-acid batteries with carbon nanotubes as an additive to the active mass. J Power Sources. 2015;296:78-85.

[85] Shapira R, Nessim GD, Zimrin T, Aurbach D. Towards

promising electrochemical technology for load leveling applications: extending cycle life of lead acid batteries by the use of carbon nano-tubes (CNTs). Energy Environ Sci. 2013;6(2):587-594.

[86] Swogger SW, Everill P, Dubey DP, Sugumaran N. Discrete carbon nanotubes increase lead acid battery charge acceptance and performance. J Power Sources. 2014;261:55-63.

[87] Banerjee A, Ziv B, Shilina Y, Levi E, Luski S, Aurbach D. Single-Wall Carbon Nanotube Doping in Lead-Acid Batteries: A New Horizon. ACS Appl Mater Interfaces. 2017;9(4):3634-3643.

[88] Geng D, Ding N, Hor TSA, Chien SW, Liu Z, Wuu D, et al. From Lithium-Oxygen to Lithium-Air Batteries: Challenges and Opportunities. Adv Energy Mater. 2016;6(9):1502164.

[89] Zhu Q-C, Du F-H, Xu S-M, Wang Z-K, Wang K-X, Chen J-S. Hydroquinone Resin Induced Carbon Nanotubes on Ni Foam As Binder-Free Cathode for Li–O2 Batteries. ACS Appl Mater Interfaces. 2016;8(6):3868-3873.

[90] Wang C, Xie Z, Zhou Z. Lithium-air batteries: Challenges coexist with opportunities. APL Materials. 2019;7(4):040701.

[91] Lee DU, Park HW, Park MG, Ismayilov V, Chen Z. Synergistic Bifunctional Catalyst Design based on Perovskite Oxide Nanoparticles and Intertwined Carbon Nanotubes for Rechargeable Zinc–Air Battery Applications. ACS Appl Mater Interfaces. 2015;7(1):902-910.

[92] Zhu S, Chen Z, Li B, Higgins D, Wang H, Li H, et al. Nitrogen-doped carbon nanotubes as air cathode catalysts in zinc-air battery. Electrochim Acta. 2011;56(14):5080-5084.

[93] Girishkumar G, Rettker M, Underhile R, Binz D, Vinodgopal K, McGinn P, et al. Single-Wall Carbon Nanotube-Based Proton Exchange Membrane Assembly for Hydrogen Fuel Cells. Langmuir. 2005;21(18):8487-8494.

[94] Tan L, Liu Z-Q, Li N, Zhang J-Y, Zhang L, Chen S. CuSe decorated carbon nanotubes as a high performance cathode catalyst for microbial fuel cells. Electrochim Acta. 2016;213:283-290.

[95] Deng D, Yu L, Chen X, Wang G, Jin L, Pan X, et al. Iron Encapsulated within Pod-like Carbon Nanotubes for Oxygen Reduction Reaction. Angew Chem Int Ed. 2013;52(1):371-375.

[96] Wen L, Liming D. Carbon Nanotube Supercapacitors. 2010.

[97] Yun YS, Yoon G, Kang K, Jin H-J. High-performance supercapacitors based on defect-engineered carbon nanotubes. Carbon. 2014;80:246-254.

[98] Frackowiak E, Jurewicz K, Delpeux S, Béguin F. Nanotubular materials for supercapacitors. J Power Sources. 2001;97-98:822-825.

[99] Futaba DN, Hata K, Yamada T, Hiraoka T, Hayamizu Y, Kakudate Y, et al. Shape-engineerable and highly densely packed single-walled carbon nanotubes and their application as super-capacitor electrodes. Nature materials. 2006;5(12):987-994.

[100] An KH, Kim WS, Park YS, Choi YC, Lee SM, Chung DC, et al. Supercapacitors Using Single-Walled Carbon Nanotube Electrodes. Adv Mater. 2001;13(7):497-500.

[101] Du C, Yeh J, Pan N. High power density supercapacitors using locally aligned carbon nanotube electrodes. Nanotechnology. 2005;16(4):350-353.

[102] Lim SH, Elim HI, Gao XY, Wee ATS, Ji W, Lee JY, et al. Electronic and optical properties of

nitrogen-doped multiwalled carbon nanotubes. Physical Review B. 2006;73(4):045402.

[103] Sun J, Huang Y, Fu C, Wang Z, Huang Y, Zhu M, et al. High-performance stretchable yarn supercapacitor based on PPy@CNTs@urethane elastic fiber core spun yarn. Nano Energy. 2016;27:230-237.

[104] Song C, Yun J, Keum K, Jeong YR, Park H, Lee H, et al. High performance wire-type supercapacitor with Ppy/CNT-ionic liquid/AuNP/carbon fiber electrode and ionic liquid based electrolyte. Carbon. 2019;144:639-648.

Chapter 7

Applications of Carbon Based Materials in Developing Advanced Energy Storage Devices

Maria Tariq, Tajamal Hussain, Adnan Mujahid, Mirza Nadeem Ahmad, Muhammad Imran Din, Azeem Intisar and Muhammad Zahid

Abstract

With the increasing pressure of population, the energy demand is growing explosively. By 2050, it is expected that the world population may reach to about 9 billion which may result in the increase of energy requirement to about 12.5 trillion watts. Due to increasing pressures of population, industries and technology, concerns to find possibilities to cope with increasing demand of energy resources, arise. Although the renewable energy resources including fossil fuels, wind, water and solar energy have been used for a long time to fulfill the energy requirements, but they need efficient conversions and storage techniques and are responsible for causing environmental pollution due to greenhouse gases as well. It is thus noteworthy to develop methods for the generation and storage of renewable energy devices that can replace the conventional energy resources to meet the requirement of energy consumption. Due to high energy demands, the sustainable energy storage devices have remained the subject of interest for scientists in the history, however, the traditional methods are not efficient enough to fulfill the energy requirements. In the present era, among other variety of advanced treatments, nano-sciences have attracted the attention of the scientists. While talking about nano-science, one cannot move on without admiring the extraordinary features of carbon nanotubes (CNTs) and other carbon based materials. CNTs are on the cutting edge of nano science research and finding enormous applications in energy storage devices. Excellent adsorption capabilities, high surface area, better electrical conductivity, high mechanical strength, corrosion resistance, high aspect ratio and good chemical and physical properties of CNTs have grabbed tremendous attention worldwide. Their charge transfer properties make them favorable for energy conversion applications. The limitation to the laboratory research on CNTs for energy storage techniques due to low specific capacitance and limited electrochemical performance can be overcome by surface functionalization using surface functional groups that can enhance their electrical and dispersion properties. In this chapter, ways CNTs employed to boost the abilities of the existing material used to store and transfer of energy have been discussed critically. Moreover, how anisotropic properties of CNTs play important role in increasing the energy storage capabilities of functional materials. It will also be discussed how various kinds of materials can be combined along CNTs to get better results.

Keywords: Energy storage, CNTs, Capacitors, Batteries

1. Introduction

With the increasing pressure of population, the energy demand is growing explosively. By 2050, it is expected that the world population may reach to about 9 billion which may result in the increase of energy requirement to about 12.5 trillion watts. Due to increasing pressures of population, industries and technology, concerns to find possibilities to cope with increasing demand of energy resources arises. Although the renewable energy resources including fossil fuels, wind, water and solar energy have been used for long time to fulfill the energy requirements, but they need efficient conversions and storage techniques and are responsible for causing environmental pollution due to greenhouse gasses as well. It is thus noteworthy to develop methods for generation and storage of renewable energy devices that can replace the conventional energy resources to meet the requirement of energy consumption.

In this chapter, we want to grab attention of the readers towards the applications of CNTs in energy storage devices. The basic principle of energy storage devices is briefly explained. Also role of carbon nanotubes as cathode and anode in different types of energy storage are discussed in this chapter.

There are two fundamental ways of storing electrochemical energy. One is the energy storage via faradic process while the other one is a non-faradic process. In the non-faradic devices, electricity is stored in electrostatic way while the faradic devices store energy electrochemically by redox reactions of active reagents. Pseudocapacitors and batteries are the examples of faradic devices while supercapacitors are the non-faradic energy storage devices.

2. Basic principle

In general, electrochemical energy storage devices involve three main steps:

- Electro sorption of ions
- Redox reaction at electrode/electrolyte interface
- Insertion of ions in to the electrodes

The energy storage devices usually store energy at the electrode/electrolyte interface in the form of accumulation of charge at the positive and negative electrodes as ions [1]. The ability of energy storage of devices is greatly affected by the electrochemical reaction that occurs at electrode electrolyte interface [1, 2].

2.1 Charging-discharging mechanism

The basic mechanism of charging and discharging of batteries as well as capacitors are discussed below.

2.1.1 Battery

Battery is composed of three main components; (i) an anode (ii) a cathode and (iii) an ionic conductor acting as an electrolyte. In order to avoid short circuit, a rigid separating medium is placed between the two electrodes (anode and cathode)

[3]. In a charged cell, movement of ions takes place from cathode to anode and reduction occurs due to ionic conduction. This electrons transportation occurs through an external circuit [4]. When a cell is discharged, oxidation occurs at anode which results in the formation of ionic species. Than these ions travel through the electrolyte and recombine at the cathode. The work is done in the process of ions transport as the ionic species produced at anode are unable to travel through the insulating electrolyte, thus they are conducted through an external circuit towards the cathode [5].

2.1.2 Capacitor

Electrochemical capacitors are divided into two main categories which are (i) electric double layer capacitor (EDLC) and (ii) pseudocapacitors. Similar to the battery, all electrochemical capacitors have a pair of electrodes which stores electrical energy [6]. An aqueous solution of acid or alkali such as that of sulfuric acid or potassium hydroxide or any other ionic liquid acts as an electrolyte [7].

There is a dielectric medium present between the electrodes of pseudocapacitors. The applied voltage produces dipoles in which electrical charges are stored. On other hand, in EDLC, electrical charges are arranged at the electrodes/electrolyte boundaries as 'electric double layer' also known as helmholtz plan [8]. The energy is delivered quickly in EDLC because of quick response of materials to the potential change and physical reactions. It is different from the behavior of battery because, the electrode potential is a continuous function of degree of charge, which is different from thermodynamic behavior of reactants of battery. It is more advantageous over battery due to its environmental friendly materials, long life span and rapid charge/discharge ability [9]. Charging-discharging pattern of the super capacitor with the time is shown in **Figure 1**. The EDLC stores charge without chemical reaction thus no heat is generated leading to high efficiency and long life. The energy stored due to fast redox reactions results in faster charging and discharging of capacitor than that of the battery. Nevertheless, due to the confined electrode surface of EDLC, the amount of energy stored in it is limited and much lower as compared to that of pseudocapacitors and batteries [10].

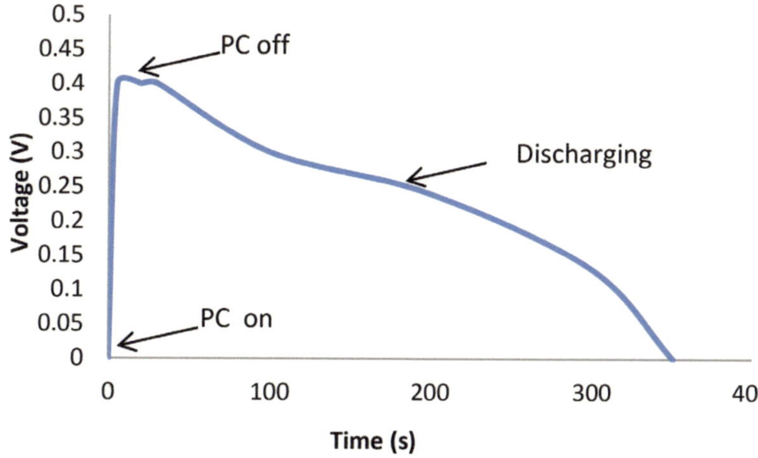

Figure 1.
Charging-discharging curve of supercapcitor.

The electric double layer can be shown in the form of equation as:

$$C = Q \backslash V = \varepsilon_o \varepsilon_r A \backslash d \quad (1)$$

Where,
C = capacitance of electrode
Q = charge transferred at potential V
ε_r = dielectric constant of electrolyte
ε_o = dielectric constant of vacuum
d = distance between electrodes
A = surface area of electrode

There are three main parameters that affect all the electrochemical energy storage devices. These include (i) specific capacitance, (ii) power, and (iii) energy density.

The total amount of electric charge that can be stored in capacitor is called the capacitance whereas the maximum amount of power that can be supplied per unit mass is called power density. The energy density can be defined as amount of energy stored per unit mass. EDLC possess lower energy densities as compared to batteries but have many advantages like high power density, faster charging and discharging, long life cycle and no change in chemical structure during charging and discharging [11].

3. CNTs for energy storage devices

Over the past many years, several advancements have been introduced in the primary conception and modification of electrode materials used for energy storage devices. Carbon-based materials, such as activated carbons (ACs), carbon nanotubes (CNTs) and graphenes have proved to be good electrode materials for energy storage devices [12, 13].

CNTs are on the cutting edge of nano science research and finding enormous applications in energy storage devices. Excellent adsorption capabilities, high surface area, better electrical conductivity, high mechanical strength, corrosion resistance, high aspect ratio and good chemical and physical properties of CNTs have grabbed tremendous attention worldwide [14, 15]. Their charge transfer properties make them favorable for energy conversion applications. The limitation to the laboratory research on CNTs for energy storage techniques due to low specific capacitance and poor electrochemical performance can be overcome by surface functionalization using surface functional groups that can enhance their electrical and dispersion properties [16]. Also the use of CNTs for energy storage devices is cheap due to easily available precursor carbon material for synthesis of CNTs. The researches on various energy storage applications of CNTs include Li-ion batteries, hydrogen storage, fuel cells and energy conversions etc.

4. Li-Ion batteries

Li-ion batteries show high energy density as compare to other rechargeable batteries. They have grabbed attention for various applications extending from electronic portable devices to electronic vehicles [17]. Among many other rechargeable batteries, LIBs have low cost, are safe for use and have least side reactions. They can offer maximum energy, high voltage, good capacity and density [18].

During the charging process, Lithium ions move from cathode to anode through an aqueous electrolyte present between the electrodes. The required driving force for this process is the chemical potential difference of Li between the electrodes. During discharging, reduction occurs at the cathode by intercalating Li-ions, while oxidation occurs at the anode simultaneously. In this way, electric current flows through external circuit to perform the required work [19, 20].

The properties of LIB such as energy density, cycle durability, rate of charging and discharging and flexibility is greatly affected by selection of suitable materials for the anode, cathode and the electrolyte [21]. The use of nanostructured materials adds many advantages over the conventional materials, such as larger contact area with electrolyte, short transport pathway for Li ions insertion and reversible Li intercalation. CNTs have been proved to be most suitable additive materials for Electrodes in LIBs and role of CNTs in LIBs is explained in the **Figure 2**. As compared to conventional LIBs, the maximum energy storing capacity of CNTs based Li-ion batteries is 1000 mAh/g (three times higher than conventional) [22].

4.1 CNTs based anode

An anode can be made of pure CNTs or composite metals, which acts as the negative electrode of the LIB during charging while cathode is composed of Li metal oxides or transition metals oxides that acts as the positive electrode of LIB in discharging. The electrochemical performance of Li ion batteries depends largely on the effective cyclic intercalation of Li ions between the electrodes. The ideal characteristics of the battery include fast charging, higher ionic storage and slow discharge [23].

Normally the metallic Lithium used as an anode in Li ion batteries causes safety issues and they have short lifetime and high cost. Carbon based materials and Li-based alloys can replace metallic Li as anode. Use of these materials reduces the activity of Li as compared to lithium metal thus results in decreasing reactivity with electrolyte, reducing the voltage of cell and improving safety. The unique structure of CNTs allows the rapid movement of Li ions through insertion and de-insertion [24, 25]. LIB anodes can be replaces by single wall carbon nanotubes as well as multiwall carbon nanotubes either by simply their deposition on a current collector or by their direct growth on a catalytically modified current collector. SWCNTs and

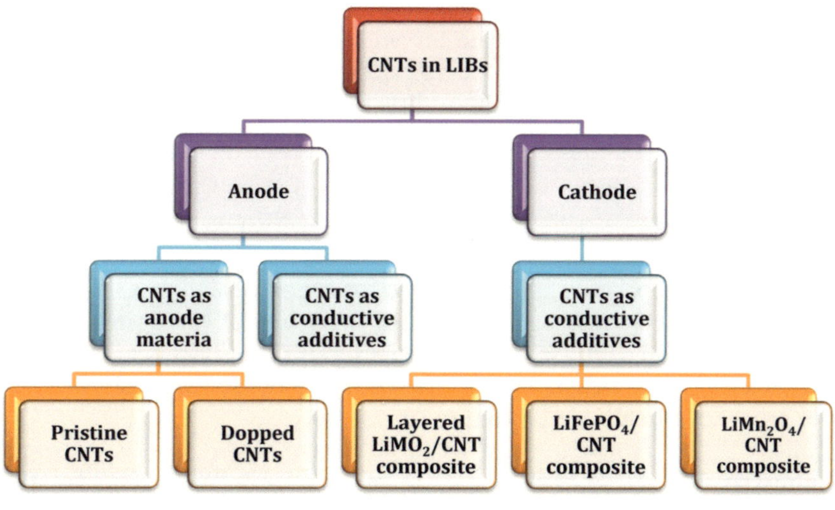

Figure 2.
Incorporation of CNTs in LIBs.

MWCNTs possess higher theoretical electrical conductivities (approximately 106 and 105S/m, respectively) and a good elastic strength (»60GPa) [26, 27].

The factors which affect the kinetics of lithium inside CNTs include radius, length, chirality and structure defects. These factors can be optimized to obtain maximum capacity results. The Li insertion capacity of carbon nanotubes in LIBs depends on chirality. The metallic CNTs show higher insertion capacities as compared to semiconductor CNTs [28].

The intercalation capacity of Li in CNT based Li-batteries is directly associated with the morphology of CNTs. Any structural defect in the morphology of CNTs affects its capacity. If there are holes in the side wall of CNTs due to defect, Li ions diffuse into them easily as compared to defect free CNTs. Li ions move randomly inside the nanotubes such that longer the length of nanotubes, slower the effective diffusion [29].

A limitation in the use of CNT anodes in LIBs is the non-reversible loss of charge after first cycle because of formation of a layer of solid electrolyte inter-phase on the CNTs. This issue can be resolved by using CNTs as conducting additives. The CNT composites with Li material have been proved to be very efficient as they resist the agglomeration as well as increase the conductivity of anode [30].

4.2 CNTs based cathode

In Li-ion batteries the active cathode material play key role in determining their performance. A variety of materials are discovered as the suitable materials for cathode of LIBs, comprising $LiCoO_2$, $LiNiO_2$, $LiMnO_2$, spinel $LiMn_2O_4$, $LiFePO_4$, $LiMPO_4$ and elemental sulfur [31–33].

The selection of appropriate cathode material greatly affect the performance of the Li-ion batteries [34]. Carbon nanotubes have been proved to be the most efficient cathode composite materials as they can reduce resistance thus increase the electrochemical performance of composite cathode. The high aspect ratio and geometry of MWCNTs provide continuous conductive network allowing efficient electron transport through material [35]. The large surface are of CNTs provides close contact with active material.

CNTs as additives for cathode materials have been reported by many researchers. Among them most widely used is the nanostructured $LiFePO_4$ with carbon nanocomposites containing monodispersed nanofibers of $LiFePO_4$ electrode [36].

For CNT based cathode, nanoparticles should have firm chemical bonds with the active materials so that CNTs act as the current-collectors for faster transport, better strength and larger surface area. CNTs can be introduced into the active material in a number of ways, including simply adding to the forerunner at the early stage of processing of active materials or by their growth in the active electrode material.

5. Super capacitors

Electrochemical capacitors, also recognized as super capacitors, are the rechargeable energy storage devices that store charge of thousands of Farads in the electrode-electrolyte interface. In contrast with other energy storage devices, super capacitors provide high power, low weight and high rate of charging-discharging [37].

Super capacitors are divided into three main types:

- Symmetric
- Asymmetric
- Hybrid

In all types of SCs, carbon is the most commonly used electrode material because they are easily available, less costly, have larger surface area and possess excellent electrical, electrochemical and mechanical properties [38].

Super capacitors are also differentiated into different types depending upon the charge storage mechanism.

- Electric double layer capacitor (Non-faradic)

- Pseudocapacitor (faradic)

The electrochemical double layer capacitor (EDLC) stores energy in a double layer of ions of electrolyte (helmholtz layer) formed on the surface of electrodes surface. The Helmholtz layer stores the charge physically. Pseudocapacitors contain electrodes of active material that store charge by faradic mechanism. Pseudocapacitors possess double the energy density as compared to EDLCs because it includes the bulk as well as the surface of the electrodes [37].

The performance of supercapacitors can be upgraded by increasing the electrode surface area or using appropriate material for electrodes. Comparison of the different features of EDLC and pseudocapictors is given in **Table 1**.

5.1 Electrode material for supercapacitors

Electrode materials play fundamental role to determine the efficiency of a supercapacitor. CNTs can be used as active materials for electrodes as well as incorporated with other additive materials. Many forms of carbon materials are proved to be effective electrode material for electrochemical capacitors. They help the ions to diffuse at the surface and also help to increase change in volume during charging-discharging.

The mostly used CNT based electrodes for supercapacitors include: [39].

- Bare CNT electrode

- Polymer/CNT composite electrode

- Metal oxide/CNT hybrid electrode

5.2 Bare CNT electrode

CNTs are frequently used as electrode material for EC capacitors due to high surface area. The capacitance of electrochemical capacitors is significantly higher

Electric double layer capacitor (EDLC)	Pseudocapacitor
Non-faradic	Faradic
Highly reversible charge–discharge	Quite reversible charge–discharge
Higher power density	Lower power density
Lower energy density	Higher energy density
20–50 mF cm^{-2}	200–500 mF cm^{-2}

Table 1.
Comparison of EDLC and pseudocapacitor.

than other capacitors; SWCNT electrodes show a capacitance of 180.0 F/g, a power density of 20.0 kW/kg and energy density of 7.0 Wh/kg [40].

CNTs can be modified for fabrication to electrode material by attachment of chemical groups through covalent bond or by wrapping the functional groups non-covalently [41]. To improve the power densities and energy, dopants are also used such as N-CNTs [42]. Furthermore, larger surface area can be obtained by oxidation. However it is difficult for bare CNTs, to obtain high energy density and power density simultaneously because of dependence of storage mechanism on physical process.

5.3 Polymer\CNT composites electrodes

Conducting polymers are grabbing the attention as supercapacitors electrode materials owing to higher specific capacitance, high conductivity in charged state, thus reduced equivalent resistance and improved power density. The randomly arranged carbon nanotubes with polymer matrix have a synergistic effect on the capacitance [43].

Among the conducting polymers, CNT composites are the most commonly used polymer composites including polyaniline [PAni] [44, 45] and polypyrrole [PPy] [46] and polythiophene (PTh) composites. We have reported in our work, electrical and thermal properties of polymethyl methacrylate CNTs composites with polyaniline-multiwalled carbon nanotubes (PANI-CNTs) as filler. Theoretically calculated percolation threshold was found to be 1.3 wt% [47]. We have also found from research that PANI had lower thermal stability than its composites with MWCNTs and Ag-MWCNTs [48].

These polymer composites exhibit several advantages like flexibility, stability, and lower cost, good electrical conductivity, more stable capacitance, and large scale production. The modification of composite due to added constituents depends upon the factors such as conductivity, accessibility and diffusion distance in electrode [49].

In one of our reported studies, Polystyrene adsorbed multi-walled carbon nanotubes incorporated polymethylmethacrylate composites have been synthesized with alleviated electrical properties. The calculated value of percolation threshold was 0.1 wt% [50].

5.4 Metal oxide\CNT composites electrodes

Metal oxides are frequently used as electrodes for electrochemical capacitors due to high densities and high strength [51]. Transition metals are more effectively used because they exhibit more than one oxidation states that results in high capacitance [52]. The faradic behavior of metal oxides depends upon the hydration properties and crystalline structure. CNTs are introduced to metal oxides so that when the composite is added to the electrode, it restricts the volume change. Among many metal oxide/CNT composites, the most widely used as electrode material is MnO_2. MnO_2 possess high theoretical capacitance, found abundantly in nature, and is environmental friendly, easily affordable and easily processed [53, 54]. Ramezani et al. reported the specific capacitance of MnO_2-CNT composites at a high scan rate of 20 mV/s, to be 180 F/g and possessed a high rate capacity [55]. Reddy et al. also reported Au doped MnO_2-CNT hybrid coaxial composites having capacitance of 68.0 F/g, energy density (4.5 Wh/kg), power density to be 33.0 kW/kg, and the cycle stability up to 1000 cycles. Effect of CNTs based metal oxide composite on the efficiency of the electrode is best explained in **Figure 3** and **Table 2** [59].

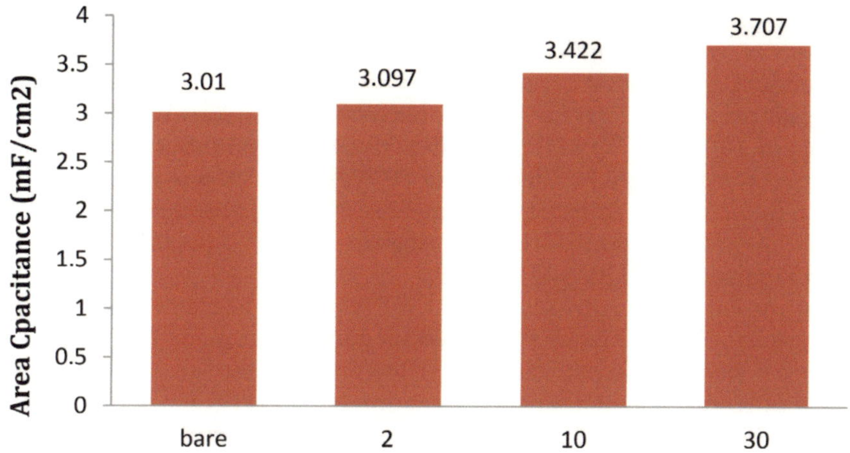

Figure 3.
Areal capacitance of CNTs fibers electrodes with different MnO_2 coating.

Sr. no.	Electrode	Specific capacitance (F/g)	Scan rate	Reference
1	MnO_2/CNT	150	20	[55]
2	MnO_2/CNT	325.5	...	[56]
3	Mn_2O_3/CNT	508	...	[57]
4	Ni(OH)2/CNF	2523	5	[58]

Table 2.
Data od CNTs based metal oxide composites as electrodes and their efficiency.

6. CNTs as flexible and separate electrodes

The energy storage devices including LIBs and super-capacitors are weighty, bulky and rigid. Therefore, they are now being replaced by the flexible storage devices due to their distinctive advantages such as less weight, flexibility and diversity of shapes etc. Therefore the flexible energy storage devices are most wanted [60].

The CNTs play an important role due to their manipulating capabilities in making flexible electrodes for flexible storage devices. CNTs play a dual role as current collector as well as active material. The CNTs thin films reduce the electrodes size and also increase flexibility and stability [61].

For the fabrication of CNTs as flexible electrodes, few aspects must be taken in account, such as young modulus of the thin film, to make sure that it may not degrade during bending or expanding. Secondly, during the charging discharging process, heat is released which may cause expansion of the material, effecting the working of the device. Thus it is also important to confirm the thermal stability of the active material [62].

6.1 CNT paper for energy storage

CNT papers having improved energy storage capabilities, have grabbed the attention for useful applications. CNT thin films are proved to possess excellent electrochemical performance due to having good conductivity, flexibility and fast heat dissipation capability [63]. With the improving technologies, CNT electrodes are being modified into CNT paper for the energy storage [64]. A number of CNT papers have been reported as electrodes for storage devices. In 2004, Morris et al. reported a free standing single walled CNTs paper electrode and its application in LIBs as initiative. This SWCNT paper is capable of showing energy of 600.0 Wh/kg and power density of nearly 3.0 kW/kg [65]. A CNT bucky-paper was invented by filtration of DWCNTs which was mechanically stable and flexible [66]. Another free flexible SWCNT paper was made by the chemical vapour deposition method, having the specific capacitance (35.0 F/g) and power density (197.3 kW/kg) [67].

The performance of energy storage of CNT paper can be enhanced by adding pseudocapacitance [68]. Xiao et al. utilized vacuum filtration method to prepare a flexible free-standing carbon nanotubes films and also used electro-chemical method in order to join redox functional groups to the CNT films [69]. The active groups containing CNT films revealed high capacitance of 150.0 mF/cm. Yang's group introduced oxygen functional groups to CNTs thin film through an acid treatment. The film showed elevated volumetric energy of approximately 200 Wh/kg and power of approximately 10 kW/kg [70].

6.2 CNT fibers

The typical weaving technology is used for making fiber shaped CNT electrodes. The fiber shaped electrodes are highly stretchable and flexible with good integration capability [71, 72]. The prime properties of electrode such as conductance, heat resistance and stability etc. are determined by the core material of the fiber. Thus, it is very important to select an appropriate material for the fiber [73].

Novel approach reported by Lu, Zan, et al. included development of super elastic hybrid CNT/graphene fiber accompanied by electro deposition of polyaniline to obtain high performing fiber supercapacitor. It was observed that the specific capacitance of prepared fiber was increased to 39% [12].

Chen, Tao, et al. invented a CNTs-based wire shaped electrode for batteries and supercapacitors and found excellent electro-chemical performance of prepared wire shaped devices with outstanding mechanical and electric properties of core CNTs [74].

6.3 CNT and polymers composites

All polymer based energy storing devices are more useful than batteries and supercapacitors due to their environmental friendly nature, flexible, low cost and versatility. In the novel approaches of flexible energy storage devices, many different ways have been used in which the electrode materials include conducting polymers [75–77] or polymers/CNTs composites [78–81].

The significance of using these polymer composite electrodes is the excellent mechanical properties and structural strength along with high tensile strength of electrode. In addition the densities of polymer-based electrodes are equivalent to that of composite electrodes [82].

Many polymer composite materials have been reported having higher electrochemical performance like Poly-pyrrole (PPY) on CoO nanowires [83], Poly-aniline (PANI) hybrid electrode [82], PPY on free CNTs bucky-paper [84] etc. Adding polymers to CNTs to form flexible composites electrode is a promising approach to obtain better electrochemical performance along with flexibility for flexible energy storage devices.

7. Flexible energy storage devices

There is a great demand of elastic energy storage devices owing to their flexibility, portability and less weight. **Figure 4** shows the importance of such flexible energy storage devices. These energy storage devices are used as wearable devices, soft electronic devices and roll up display [85, 86]. In order to achieve flexible energy storage devices, the main challenge is to select appropriate material having high capacity and conductivity. There are two main types of elastic energy storage devices:

- Flexible LIBs
- Flexible supercapacitors

7.1 Flexible Li-ion batteries

In order to design portable electronics such as smart cards, wireless sensor, wearable devices, roll up displays etc. flexible Li ion batteries are required which have high energy density and excellent rate capabilities [87]. Flexible batteries have been developed by many routes including cellulose based batteries [88], polymer batteries [89], soft packing batteries [90], and paper based batteries [91, 92]. The performance of flexible batteries highly depends upon the type of electrode material thus a soft flexible nanostructured material is highly recommended to construct a flexible battery. Carbon nanotubes, owing to their unique properties like extremely flexible and highly conductive, take their top priority to be used as electrode material for flexible batteries [93].

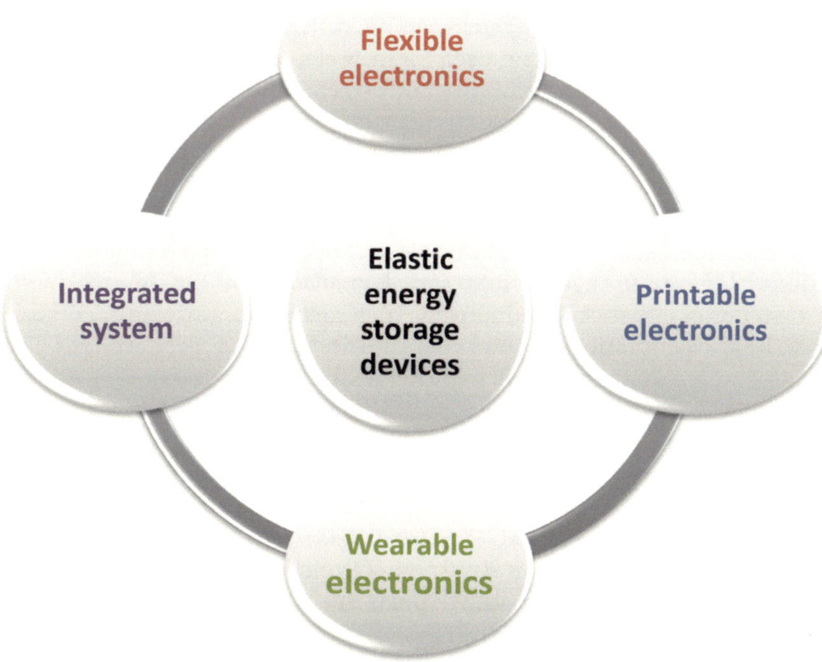

Figure 4.
Applications of flexible energy storage devices.

Ajayan et al. reported porous cellulose paper having CNTs embedded on it used as electrode. The paper was capable of bending, twisting and rolling to any degree [63]. Ren Jingn et al. used MWCNT/LiO_2 as electrodes to form a malleable wire-shaped Li-ion battery. The battery showed the power-density of 880 W/kg and energy-density of 27 Wh/kg. The prepared wire-shaped batteries were fabricated into low weight, flexible and malleable battery textile to check their application [94].

Fang et al. developed a lithium sulfur battery by twisting a fibrous cathode fabricated by aligned CNTs coated with sulfur and an anode of Li wire. The composite cathode displayed capacity of 1051 mAh/g versus sulfur which retained 600 mAh/g after 100 running cycles, showing good cycling performance [95].

7.2 Flexible supercapaitors

In the modern era, transportable electronic devices including mobiles, wearable electronics and light weight elastic electronic devices are of great demand. While talking about portable energy storage devices, one cannot ignore supercapacitors. Supercapacitors are having applications in every electronic device because of higher specific capacitance and power density [96–98]. Therefore, flexible supercapacitors are always preferred for elastic electronic devices. CNTs are proved to be excellent electrode material for flexible supercapacitors owing to their high aspect ratio, high conductance and porosity [99, 100].

Wang, Q. et al. reported synthesis of strong flexible CNT-MnO_2 nanosheets with excellent capacitance for flexible supercapacitor [101]. In another approach, reduced graphene-oxide and carbon nanotubes were developed as electrodes for flexible supercapacitors. The addition of CNTs provided a dense structure having mesopores in hybrid fiber. The electrode exhibits high tensile strength, high conductance and capacitance of 354.9 F/cm^3 [102]. CuO/MWCNTs nanocomposites were synthesized which showed the specific-capacitance of 452.8 F/g and the scan ate of 10 mV/s [103].

Niu et al. prepared stretchable buckled SWCNT films combined with polydimethylsiloxane (PDMS) and used them as electrode fo flexible supercapacitor [104] it showed maximum flexibility and strechability.

8. Conclusion

CNTs are on the cutting edge of nano science research and finding enormous applications in energy storage devices. Excellent adsorption capabilities, high surface area, better electrical conductivity, high mechanical strength, corrosion resistance, high aspect ratio and good chemical and physical properties of CNTs have grabbed tremendous attention worldwide. Among energy storage devices, Li ion batteries, electric double layer capacitors and pseudocapacitors are more commonly used. In Li-ion batteries CNTs are use as cathodes as well as anodes. It is observed that as compared to conventional LIBs, the maximum energy storing capacity of CNTs based Li-ion batteries is 1000 mAh/g i.e. three times higher than conventional. In case of supercapacitors, CNTs based electrodes include bare CNTs, polymer/CNTs electrodes and metal oxide/CNTs electrodes. Carbon nanotubes based flexible electrodes have become popular due to their distinctive advantages such as less weight, flexibility and diversity of shapes etc. Flexible energy storage devices such as flexible lithium ion batteries and flexible super capacitors are used as wearable devices, soft electronic devices and roll up display. In order to achieve flexible energy storage devices, the main challenge of selecting appropriate material having high capacity and conductivity can be achieved by using carbon nanotubes.

Author details

Maria Tariq[1], Tajamal Hussain[1*], Adnan Mujahid[1], Mirza Nadeem Ahmad[2], Muhammad Imran Din[1], Azeem Intisar[1] and Muhammad Zahid[3]

1 Institute of Chemistry, University of the Punjab, Lahore, Pakistan

2 Department of Applied Chemistry, Govt College University Faisalabad, Pakistan

3 Department of Chemistry, Agriculture University of Faisalabad, Faisalabad, Pakistan

*Address all correspondence to: tajamalhussain.chem@pu.edu.pk

References

[1] Liu, J., et al., *Advanced energy storage devices: basic principles, analytical methods, and rational materials design.* Advanced science, 2018. **5**(1): p. 1700322.

[2] Sumboja, A., et al., *Electrochemical energy storage devices for wearable technology: a rationale for materials selection and cell design.* Chemical Society Reviews, 2018. **47**(15): p. 5919-5945.

[3] Mehtab, T., et al., *Metal-organic frameworks for energy storage devices: batteries and supercapacitors.* Journal of Energy Storage, 2019. **21**: p. 632-646.

[4] Li, J., et al., *Studies on the cycle life of commercial lithium ion batteries during rapid charge–discharge cycling.* Journal of Power Sources, 2001. **102**(1-2): p. 294-301.

[5] Gogotsi, Y. and R.M. Penner, *Energy storage in nanomaterials–capacitive, pseudocapacitive, or battery-like?* 2018, ACS Publications.

[6] Eftekhari, A., *Metrics for fast supercapacitors as energy storage devices*. 2018, ACS Publications.

[7] Merlet, C., et al., *On the molecular origin of supercapacitance in nanoporous carbon electrodes.* Nature materials, 2012. **11**(4): p. 306-310.

[8] Boota, M., et al., *Organic-inorganic all-pseudocapacitive asymmetric energy storage devices.* Nano Energy, 2019. **65**: p. 104022.

[9] Miller, E.E., Y. Hua, and F.H. Tezel, *Materials for energy storage: Review of electrode materials and methods of increasing capacitance for supercapacitors.* Journal of Energy Storage, 2018. **20**: p. 30-40.

[10] Zhou, Y., et al., *Ultrahigh-Areal-Capacitance Flexible Supercapacitor Electrodes Enabled by Conformal P3MT on Horizontally Aligned Carbon-Nanotube Arrays.* Advanced Materials, 2019. **31**(30): p. 1901916.

[11] Allagui, A., et al., *Capacitive behavior and stored energy in supercapacitors at power line frequencies.* Journal of Power Sources, 2018. **390**: p. 142-147.

[12] Lu, Z., et al., *Superelastic hybrid CNT/graphene fibers for wearable energy storage.* Advanced Energy Materials, 2018. **8**(8): p. 1702047.

[13] Wen, L., F. Li, and H.M. Cheng, *Carbon nanotubes and graphene for flexible electrochemical energy storage: from materials to devices.* Advanced Materials, 2016. **28**(22): p. 4306-4337.

[14] De Volder, M.F., et al., *Carbon nanotubes: present and future commercial applications.* science, 2013. **339**(6119): p. 535-539.

[15] Soni, S.K., B. Thomas, and V.R. Kar, *A Comprehensive Review on CNTs and CNT-Reinforced Composites: Syntheses, Characteristics and Applications.* Materials Today Communications, 2020: p. 101546.

[16] Jia, X. and F. Wei, *Advances in production and applications of carbon nanotubes*, in *Single-Walled Carbon Nanotubes*. 2019, Springer. p. 299-333.

[17] Li, H., *Practical evaluation of Li-ion batteries.* Joule, 2019. **3**(4): p. 911-914.

[18] El Kharbachi, A., et al., *Exploits, advances and challenges benefiting beyond Li-ion battery technologies.* Journal of Alloys and Compounds, 2020. **817**: p. 153261.

[19] Liu, S., et al., *Deep-discharging li-ion battery state of charge estimation using a partial adaptive forgetting factors least*

square method. IEEE Access, 2019. **7**: p. 47339-47352.

[20] He, Y., et al., *A new model for State-of-Charge (SOC) estimation for high-power Li-ion batteries.* Applied Energy, 2013. **101**: p. 808-814.

[21] Lin, C., et al., *Li 4 Ti 5 O 12-based anode materials with low working potentials, high rate capabilities and high cyclability for high-power lithium-ion batteries: A synergistic effect of doping, incorporating a conductive phase and reducing the particle size.* Journal of Materials Chemistry A, 2014. **2**(26): p. 9982-9993.

[22] Liu, J., *Addressing the grand challenges in energy storage.* Advanced Functional Materials, 2013. **23**(8): p. 924-928.

[23] Liu, X.-M., et al., *Carbon nanotube (CNT)-based composites as electrode material for rechargeable Li-ion batteries: a review.* Composites Science and Technology, 2012. **72**(2): p. 121-144.

[24] Chen, Y., et al., *Hollow carbon-nanotube/carbon-nanofiber hybrid anodes for Li-ion batteries.* Journal of the American Chemical Society, 2013. **135**(44): p. 16280-16283.

[25] Chen, L., et al., *Porous graphitic carbon nanosheets as a high-rate anode material for lithium-ion batteries.* ACS applied materials & interfaces, 2013. **5**(19): p. 9537-9545.

[26] Goriparti, S., et al., *Review on recent progress of nanostructured anode materials for Li-ion batteries.* Journal of power sources, 2014. **257**: p. 421-443.

[27] Kang, C., et al., *3-dimensional carbon nanotube for Li-ion battery anode.* Journal of Power Sources, 2012. **219**: p. 364-370.

[28] Kawasaki, S., et al., *Metallic and semiconducting single-walled carbon nanotubes as the anode material of Li ion secondary battery.* Materials Letters, 2008. **62**(17-18): p. 2917-2920.

[29] Bhatt, M.D. and C. O'Dwyer, *Recent progress in theoretical and computational investigations of Li-ion battery materials and electrolytes.* Physical Chemistry Chemical Physics, 2015. **17**(7): p. 4799-4844.

[30] Wang, X.X., et al., *Preparation of short carbon nanotubes and application as an electrode material in Li-ion batteries.* Advanced Functional Materials, 2007. **17**(17): p. 3613-3618.

[31] Park, K., et al., *Electrochemical nature of the cathode interface for a solid-state lithium-ion battery: interface between LiCoO2 and garnet-Li7La3Zr2O12.* Chemistry of Materials, 2016. **28**(21): p. 8051-8059.

[32] Xiong, X., et al., *Role of V2O5 coating on LiNiO2-based materials for lithium ion battery.* Journal of Power Sources, 2014. **245**: p. 183-193.

[33] Yu, H. and H. Zhou, *High-energy cathode materials (Li2MnO3–LiMO2) for lithium-ion batteries.* The journal of physical chemistry letters, 2013. **4**(8): p. 1268-1280.

[34] Li, Q., et al., *Conjugated carbonyl polymer-based flexible cathode for superior lithium-organic batteries.* ACS applied materials & interfaces, 2019. **11**(32): p. 28801-28808.

[35] Diao, G., et al., *Nickel and cobalt effect on properties of MWCNT-based anode for Li-ion batteries.* Applied Nanoscience, 2020: p. 1-7.

[36] Sides, C.R., et al., *A high-rate, nanocomposite LiFePO4/ carbon cathode.* Electrochemical and Solid State Letters, 2005. **8**(9): p. A484.

[37] Sarno, M., *Nanotechnology in energy storage: the supercapacitors*, in *Studies in*

Surface Science and Catalysis. 2019, Elsevier. p. 431-458.

[38] Li, J., *Review of electrochemical capacitors based on carbon nanotubes and graphene.* Graphene, 2012. **1**(01): p. 1.

[39] Kumar, S., et al., *Carbon nanotubes: A potential material for energy conversion and storage.* Progress in energy and combustion science, 2018. **64**: p. 219-253.

[40] Tashima, D., et al., *Space charge distributions of an electric double layer capacitor with carbon nanotubes electrode.* Thin Solid Films, 2007. **515**(9): p. 4234-4239.

[41] Hiraoka, T., et al., *Compact and Light Supercapacitor Electrodes from a Surface-Only Solid by Opened Carbon Nanotubes with 2 200 m2 g− 1 Surface Area.* Advanced Functional Materials, 2010. **20**(3): p. 422-428.

[42] Sevilla, M., et al., *Surface modification of CNTs with N-doped carbon: an effective way of enhancing their performance in supercapacitors.* ACS Sustainable Chemistry & Engineering, 2014. **2**(4): p. 1049-1055.

[43] Magu, T.O., et al., *A review on conducting polymers-based composites for energy storage application.* Journal of Chemical Reviews, 2019. **1**(1, pp. 1-77.): p. 19-34.

[44] Dong, B., et al., *Preparation and electrochemical characterization of polyaniline/multi-walled carbon nanotubes composites for supercapacitor.* Materials Science and Engineering: B, 2007. **143**(1-3): p. 7-13.

[45] Yazdi, M.K., et al., *PANI-CNT nanocomposites*, in *Fundamentals and Emerging Applications of Polyaniline*. 2019, Elsevier. p. 143-163.

[46] An, K.H., et al., *High-capacitance supercapacitor using a nanocomposite electrode of single-walled carbon nanotube and polypyrrole.* Journal of the Electrochemical Society, 2002. **149**(8): p. A1058.

[47] Bashir, F., et al., *Tailoring electrical and thermal properties of polymethyl methacrylate-carbon nanotubes composites through polyaniline and dodecyl benzene sulphonic acid impregnation.* Polymer Composites, 2018. **39**(S2): p. E1052-E1059.

[48] Hussain, T., et al., *Polyaniline/silver decorated-MWCNT composites with enhanced electrical and thermal properties.* Polymer Composites, 2018. **39**(S3): p. E1346-E1353.

[49] Wei, C., et al., *Polymer composites with functionalized carbon nanotube and graphene*, in *Polymer Composites with Functionalized Nanoparticles*. 2019, Elsevier. p. 211-248.

[50] Hussain, T., et al., *Polystyrene adsorbed multi-walled carbon nanotubes incorporated polymethylmethacrylate composites with modified percolation phenomena.* MRS Advances, 2018. **3**(1): p. 25-30.

[51] Zhi, M., et al., *Nanostructured carbon–metal oxide composite electrodes for supercapacitors: a review.* Nanoscale, 2013. **5**(1): p. 72-88.

[52] Lee, T.H., et al., *High energy density and enhanced stability of asymmetric supercapacitors with mesoporous MnO2@ CNT and nanodot MoO3@ CNT free-standing films.* Energy Storage Materials, 2018. **12**: p. 223-231.

[53] Wu, P., et al., *Synthesis and characterization of self-standing and highly flexible δ-MnO2@ CNTs/CNTs composite films for direct use of supercapacitor electrodes.* ACS applied materials & interfaces, 2016. **8**(36): p. 23721-23728.

[54] Yang, P., et al., *Low-cost high-performance solid-state asymmetric*

supercapacitors based on MnO2 nanowires and Fe2O3 nanotubes. Nano letters, 2014. **14**(2): p. 731-736.

[55] Ramezani, M., M. Fathi, and F. Mahboubi, *Facile synthesis of ternary MnO2/graphene nanosheets/carbon nanotubes composites with high rate capability for supercapacitor applications.* Electrochimica Acta, 2015. **174**: p. 345-355.

[56] Huang, M., et al., *Layered manganese oxides-decorated and nickel foam-supported carbon nanotubes as advanced binder-free supercapacitor electrodes.* Journal of Power Sources, 2014. **269**: p. 760-767.

[57] Zhou, R., et al., *High-performance supercapacitors using a nanoporous current collector made from super-aligned carbon nanotubes.* Nanotechnology, 2010. **21**(34): p. 345701.

[58] Zhang, L., et al., *Flexible hybrid membranes with Ni (OH) 2 nanoplatelets vertically grown on electrospun carbon nanofibers for high-performance supercapacitors.* ACS applied materials & interfaces, 2015. 7(40): p. 22669-22677.

[59] Reddy, A.L.M., et al., *Multisegmented Au-MnO2/carbon nanotube hybrid coaxial arrays for high-power supercapacitor applications.* The Journal of Physical Chemistry C, 2010. **114**(1): p. 658-663.

[60] Gwon, H., et al., *Recent progress on flexible lithium rechargeable batteries.* Energy & Environmental Science, 2014. 7(2): p. 538-551.

[61] Xiao, X., et al., *Freestanding mesoporous VN/CNT hybrid electrodes for flexible all-solid-state supercapacitors.* Advanced Materials, 2013. **25**(36): p. 5091-5097.

[62] Utsunomiya, T., et al., *Self-discharge behavior and its temperature dependence of carbon electrodes in lithium-ion batteries.* Journal of Power Sources, 2011. **196**(20): p. 8598-8603.

[63] Pushparaj, V.L., et al., *Flexible energy storage devices based on nanocomposite paper.* Proceedings of the National Academy of Sciences, 2007. **104**(34): p. 13574-13577.

[64] Hu, L., et al., *Highly conductive paper for energy-storage devices.* Proceedings of the National Academy of Sciences, 2009. **106**(51): p. 21490-21494.

[65] Morris, R.S., et al., *High-energy, rechargeable Li-ion battery based on carbon nanotube technology.* Journal of Power Sources, 2004. **138**(1-2): p. 277-280.

[66] Endo, M., et al., *'Buckypaper'from coaxial nanotubes.* Nature, 2005. **433**(7025): p. 476-476.

[67] Niu, Z., et al., *Compact-designed supercapacitors using free-standing single-walled carbon nanotube films.* Energy & Environmental Science, 2011. **4**(4): p. 1440-1446.

[68] Yao, B., et al., *Paper-based electrodes for flexible energy storage devices.* Advanced Science, 2017. **4**(7): p. 1700107.

[69] Cheng, Y., et al., *Flexible and cross-linked N-doped carbon nanofiber network for high performance freestanding supercapacitor electrode.* Nano energy, 2015. **15**: p. 66-74.

[70] Lee, S.W., et al., *Self-standing positive electrodes of oxidized few-walled carbon nanotubes for light-weight and high-power lithium batteries.* Energy & Environmental Science, 2012. **5**(1): p. 5437-5444.

[71] Zhang, Y., et al., *High-performance lithium–air battery with a coaxial-fiber architecture.* Angewandte Chemie International Edition, 2016. **55**(14): p. 4487-4491.

[72] Wang, B., et al., *Fabricating continuous supercapacitor fibers with high performances by integrating all building materials and steps into one process.* Advanced Materials, 2015. **27**(47): p. 7854-7860.

[73] Pan, S., et al., *Wearable solar cells by stacking textile electrodes.* Angewandte Chemie, 2014. **126**(24): p. 6224-6228.

[74] Chen, T., et al., *Nitrogen-Doped Carbon Nanotube Composite Fiber with a Core–Sheath Structure for Novel Electrodes.* Advanced Materials, 2011. **23**(40): p. 4620-4625.

[75] Wang, J.-Z., et al., *Highly flexible and bendable free-standing thin film polymer for battery application.* Materials Letters, 2009. **63**(27): p. 2352-2354.

[76] Wang, C., et al., *Functionalised polyterthiophenes as anode materials in polymer/polymer batteries.* Synthetic metals, 2010. **160**(1-2): p. 76-82.

[77] Mihranyan, A., et al., *A novel high specific surface area conducting paper material composed of polypyrrole and Cladophora cellulose.* The Journal of Physical Chemistry B, 2008. **112**(39): p. 12249-12255.

[78] Meng, C., C. Liu, and S. Fan, *Flexible carbon nanotube/polyaniline paper-like films and their enhanced electrochemical properties.* Electrochemistry communications, 2009. **11**(1): p. 186-189.

[79] Xiao, Q. and X. Zhou, *The study of multiwalled carbon nanotube deposited with conducting polymer for supercapacitor.* Electrochimica Acta, 2003. **48**(5): p. 575-580.

[80] Frackowiak, E., et al., *Supercapacitors based on conducting polymers/nanotubes composites.* Journal of Power Sources, 2006. **153**(2): p. 413-418.

[81] Wang, J., et al., *Highly-flexible fibre battery incorporating polypyrrole cathode and carbon nanotubes anode.* Journal of power sources, 2006. **161**(2): p. 1458-1462.

[82] Patil, D.S., et al., *Polyaniline based electrodes for electrochemical supercapacitor: Synergistic effect of silver, activated carbon and polyaniline.* Journal of Electroanalytical Chemistry, 2014. **724**: p. 21-28.

[83] Zhou, C., et al., *Construction of high-capacitance 3D CoO@ polypyrrole nanowire array electrode for aqueous asymmetric supercapacitor.* Nano letters, 2013. **13**(5): p. 2078-2085.

[84] Che, J., P. Chen, and M.B. Chan-Park, *High-strength carbon nanotube buckypaper composites as applied to free-standing electrodes for supercapacitors.* Journal of Materials Chemistry A, 2013. **1**(12): p. 4057-4066.

[85] Gates, B.D., *Flexible electronics.* Science, 2009. **323**(5921): p. 1566-1567.

[86] Bauer, S., *Flexible electronics: sophisticated skin.* Nature materials, 2013. **12**(10): p. 871-872.

[87] Liu, J. and X.W. Liu, *Two-dimensional nanoarchitectures for lithium storage.* Advanced materials, 2012. **24**(30): p. 4097-4111.

[88] Jabbour, L., et al., *Cellulose-based Li-ion batteries: a review.* Cellulose, 2013. **20**(4): p. 1523-1545.

[89] Nyholm, L., et al., *Toward flexible polymer and paper-based energy storage devices.* Advanced Materials, 2011. **23**(33): p. 3751-3769.

[90] Choi, K.H., et al., *Thin, deformable, and safety-reinforced plastic crystal polymer electrolytes for high-performance flexible lithium-ion batteries.* Advanced Functional Materials, 2014. **24**(1): p. 44-52.

[91] Li, N., et al., *Flexible graphene-based lithium ion batteries with ultrafast charge and discharge rates.* Proceedings of the National Academy of Sciences, 2012. **109**(43): p. 17360-17365.

[92] Zhu, H., et al., *Tin anode for sodium-ion batteries using natural wood fiber as a mechanical buffer and electrolyte reservoir.* Nano letters, 2013. **13**(7): p. 3093-3100.

[93] Liu, L., W. Ma, and Z. Zhang, *Macroscopic carbon nanotube assemblies: preparation, properties, and potential applications.* Small, 2011. **7**(11): p. 1504-1520.

[94] Ren, J., et al., *Elastic and wearable wire-shaped lithium-ion battery with high electrochemical performance.* Angewandte Chemie, 2014. **126**(30): p. 7998-8003.

[95] Fang, X., et al., *A cable-shaped lithium sulfur battery.* Advanced materials, 2016. **28**(3): p. 491-496.

[96] Lu, X., et al., *H-TiO2@ MnO2//H-TiO2@ C core–shell nanowires for high performance and flexible asymmetric supercapacitors.* Advanced materials, 2013. **25**(2): p. 267-272.

[97] Zhou, S., et al., *Cellulose Nanofiber@ Conductive Metal–Organic Frameworks for High-Performance Flexible Supercapacitors.* ACS nano, 2019. **13**(8): p. 9578-9586.

[98] Han, Y. and L. Dai, *Conducting polymers for flexible supercapacitors.* Macromolecular Chemistry and Physics, 2019. **220**(3): p. 1800355.

[99] Liu, L., Z. Niu, and J. Chen, *Flexible supercapacitors based on carbon nanotubes.* Chinese Chemical Letters, 2018. **29**(4): p. 571-581.

[100] Zhu, S., J. Ni, and Y. Li, *Carbon nanotube-based electrodes for flexible supercapacitors.* NANO RESEARCH, 2020.

[101] Wang, Q., et al., *Flexible supercapacitors based on carbon nanotube-MnO2 nanocomposite film electrode.* Chemical Engineering Journal, 2019. **371**: p. 145-153.

[102] Xu, T., et al., *Reduced graphene oxide/carbon nanotube hybrid fibers with narrowly distributed mesopores for flexible supercapacitors with high volumetric capacitances and satisfactory durability.* Carbon, 2019. **152**: p. 134-143.

[103] Paulose, R. and M. Raja, *CuO nanoparticles/multi-walled carbon nanotubes (MWCNTs) nanocomposites for flexible supercapacitors.* Journal of nanoscience and nanotechnology, 2019. **19**(12): p. 8151-8156.

[104] Niu, Z., et al., *Highly stretchable, integrated supercapacitors based on single-walled carbon nanotube films with continuous reticulate architecture.* Advanced Materials, 2013. **25**(7): p. 1058-1064.

Section 5

Error-Resilience in Multi-Valued CNTFET Logic

Chapter 8

Fault Tolerance in Carbon Nanotube Transistors Based Multi Valued Logic

Gopalakrishnan Sundararajan

Abstract

This Chapter presents a solution for fault-tolerance in Multi-Valued Logic (MVL) circuits comprised of Carbon Nano-Tube Field Effect Transistors (CNTFET). This chapter reviews basic primitives of MVL and describes ternary implementations of CNTFET circuits. Finally, this chapter describes a method for error correction called Restorative Feedback (RFB). The RFB method is a variant of Triple-Modular Redundancy (TMR) that utilizes the fault masking capabilities of the Muller C element to provide added protection against noisy transient faults. Fault tolerant properties of Muller C element is discussed and error correction capability of RFB method is demonstrated in detail.

Keywords: CNTFET, Multi-Valued Logic, Fault Tolerance, TMR

1. Introduction

Moore's law along with the Dennard scaling has been the key driving factor that has enabled steady progress in the semiconductor industry over the last three decades. However, continuous scaling of Complementary Metal Oxide Semiconductor (CMOS) transistor into the sub-nanometer regime faces severe challenges due to short channel effects, exponentially rising leakage currents and increased variations in manufacturing process. To combat these challenges, numerous approaches have been explored. One of the promising approaches is to look for alternatives transistor structures that could potentially overcome these challenges. In this regard, Quantum-dot Cellular Automata (QCA), Carbon Nano-Tube Field Effect Transistor (CNTFET) and Single Electron Transistor (SET) are proposed to replace or supplement CMOS technology [1, 2].

From circuits and systems design perspective, several approaches are explored to reduce die area and lower energy consumption. One of the approaches is to use Multi-Valued Logic (MVL) design. MVL circuit design has been explored in CMOS technology for last few decades using circuit styles like voltage-mode CMOS logic (VMCL), I^2L logic and current-mode CMOS logic (CMCL) [3, 4]. Binary logic packs two logic values between available voltage levels, whereas MVL packs more than two logic values between the available voltage levels. Ternary logic is simplest form of MVL. Due to reduced noise margins, MVL logic design is less reliable and more prone to defects. Significant efforts have been directed towards defect and fault tolerance for binary logic and less effort has been directed towards defect and fault tolerant techniques for MVL.

In digital design, transistor is used as switch that transitions between logic states. The switch that controls this transition is the threshold voltage of the transistor. Threshold voltage is a manufacturing process parameter that is set to a unique value during manufacturing. To realize CMOS MVL circuits, multi-threshold transistors are deployed. In CMOS, multi-threshold transistors are realized by applying body bias voltages that exploit body effect to alter threshold voltages [5]. However, CNTFETs are fundamentally different that they allow realization of multi-threshold transistors by tuning a few process parameters. This property has been effectively exploited to realize various forms of ternary logic circuits using CNTFET [6, 7].

Despite several advantages, there are significant issues with the reliable realization of CNTFET circuits. There are fundamental limitations that are specific to Carbon Nanotubes (CNT) that pose major challenges [8]. CNTs are graphene sheets rolled into tubes [9]. Multiple CNTs are deployed in the channel region to provide the required drive currents needed for reliable operation [10]. These CNTs can be either metallic or semiconducting depending on the arrangement of the carbon atoms in the tube. Metallic tubes can result in circuit malfunctioning due to source-drain shorts [8]. It is also not possible to guarantee perfect positioning and alignment of these CNTs in large CNTFET circuits [8]. To harness the potential benefits of CNTFET technology, variation aware defect and fault tolerant techniques are needed for reliable operation of CNTFET MVL circuits.

In this chapter, we present MVL realization of CNTFET circuits and discuss techniques for defect and fault tolerance in MVL CNTFET circuits. The rest of this chapter is organized as follows: Section 2 discusses basic logic primitives that are needed for understanding MVL. Section 3 describes the CNTFET transistor and its operation. Section 4 provides a detail description of various circuit styles that have been proposed to realize MVL CNTFET circuits. Section 5 describes variation in CNTFET devices. Section 6 describes a technique for fault tolerance in MVL CNTFET circuits.

2. MVL basics

Raychowdhury et al. detail and discuss the logic primitives that are needed for understanding MVL [6]. Let us consider an r-valued n-variable function $f(X)$, where $X = x_1, x_2, x_3, \dots x_n$ and each x_i can take up values from $R = 0, 1, 2, \dots, r-1$.

The mapping of $f(X)$ is $f : R^n - > R$. There are r^{r^n} different functions that are possible in set f. Ternary logic gates that implement each of these functions would be called primitive gates. There are three main primitives that are useful for understanding MVL: complement, min and tsum. Complement operator is equivalent to a ternary inverter. Min operator is equivalent to a ternary and gate. The complement of the min operator is equivalent to ternary nand gate. Tsum operator is equivalent to a ternary or gate and the complement of tsum operator is a ternary nor gate. Complement of a logic value l is defined as:

$$\hat{l} = (r-1) - l. \tag{1}$$

Consider r = 3. The complement set is given by **Table 1** [9]. A min operator is defined as

$$\min(a_1, a_2, a_3, \dots a_n) = a_1 \cdot a_2 \cdot a_3 \cdot \dots a_n \tag{2}$$

Input l	Output $\hat{l}$
0	2
1	1
2	0

Table 1.
Truth table for complement operator [9].

Input a	Input b	Output c_{nand}	Output c_{nor}
0	0	2	2
1	0	2	1
2	0	2	0
0	1	2	1
1	1	1	1
2	1	1	0
0	2	2	0
1	2	1	0
2	2	0	0

Table 2.
Ternary gates truth table [7].

Ternary nand operator can be defined as

$$c_{nand} = \overline{\min(a_1, a_2, a_3, \dots a_n)} = \overline{a_1 \cdot a_2 \cdot a_3 \cdot \dots a_n} \tag{3}$$

Here is an example of min operator: $min\,(1, 2, 3) = 1$. A tsum operator is defined as

$$\text{tsum}(a_1, a_2, a_3, \dots a_n) = a_1 \oplus a_2 \oplus a_3 \dots a_n \tag{4}$$

$$\text{tsum}(a_1, a_2, a_3, \dots a_n) = \min(a_1 + a_2 + a_3 + \dots + a_n, r - 1) \tag{5}$$

Ternary nor operator can be defined as:

$$c_{nor} = \overline{\text{tsum}(a_1, a_2, a_3, \dots a_n)} = \overline{\min(a_1 + a_2 + a_3 + \dots + a_n, r - 1)} \tag{6}$$

where belongs to the set R. Consider $r = 3$, $tsum(1, 1, 0) = min\,(1 + 1 + 0, 2) = 2$. **Table 2** describes the truth table for ternary nand and ternary nor [7].

3. CNTFET transistor

Figure 1 shows the cross-section of a CNTFET transistor. Similar to MOSFETs, CNTFETs have four terminals: drain, gate, source and substrate [11]. CNTFET transistors are constructed by replacing the silicon channel in CMOS with carbon nanotubes (CNT). CNTs are sheets of Graphene rolled into tubes. Depending on the direction in which the sheets are rolled in the channel, CNTs can be either metallic or semi-conducting. This property of CNT being metallic or non-metallic depending on their rolled direction is termed as chirality [7]. In CMOS transistors, drive current depends on the channel width [5]. However, in the case of CNTFET the drive currents during conduction state depends on the number of CNTs in the

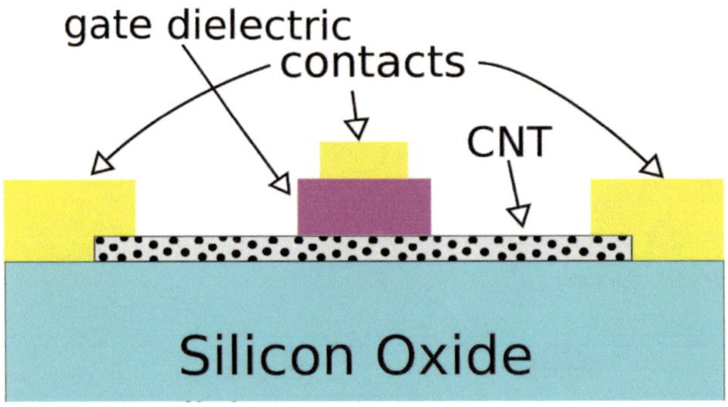

Figure 1.
Side view of a CNTFET transistor.

channel along with gate length, chirality and pitch distance [12]. The V_{th} of the intrinsic CNT channel is given by

$$V_{\text{th}} \approx \frac{Eg}{2e} = \frac{\sqrt{3}}{3} \frac{aV\pi}{eD_{\text{CNT}}} \tag{7}$$

where $a = 2.49 \times 10^{-10}$, m is the carbon to carbon atom distance, $V_{\pi} = 3.033$ ev is the carbon $\pi - \pi$ bond energy in the tight bonding model, e is the unit electron charge and D_{CNT} is the tube diameter [13]. D_{CNT} is calculated using the following Equation [13].

$$D_{\text{CNT}} = \frac{\sqrt{3}a_0}{\pi} \sqrt{n^2 + m^2 + nm} \tag{8}$$

where a_0 = 0.142 nm is the inter-atomic distance between each carbon atom and its neighbor and (n,m) is called the chirality vector that describes the structure of a carbon nanotube, n is the number of non-metallic tubes and m is the number of metallic tubes present in the channel. The diameter D_{CNT} used in the Stanford CNTFET model is $1.487nm$ [10]. The model assumes that the CNT synthesis process yields zero metallic tubes. The value of n from equation (8) amounts to 19. This corresponds to a chirality vector of (19, 0) and threshold voltage of $0.293V$ based on equation (7). Combination of equations (7) and (8), provides the following insight: Threshold voltage of the CNTFET can be modified by changing the number of non-metallic tubes assuming that CNT synthesis process yields zero metallic tubes. The ratio of threshold voltages of two CNTFETs is inversely proportional to the number of non-metallic tubes present in the CNTFET.

$$\frac{V_{\text{th1}}}{V_{\text{th2}}} = \frac{D_{\text{CNT2}}}{D_{\text{CNT1}}} = \frac{n_2}{n_1}. \tag{9}$$

This property has been effectively exploited to design MVL circuits using CNTFET. In practice, multi-diameter CNTFET is realized by CNT synthesis techniques that can fabricate CNTs with desired chirality [14]. Also, post-processing techniques for adjusting the threshold voltage of multiple tube CNTFET have also been demonstrated [15].

4. Circuit realizations of ternary logic gates

This section discusses two circuit implementations of CNTFET ternary inverters along with a circuit implementation of a CNTFET ternary Muller C element. First circuit is a resistor based ternary inverter that was proposed by Raychowdhury et al. [9]. Second circuit is a ternary inverter using static complementary circuit style that was proposed by Lin et al. [7]. We will also discuss the circuit implementation and operation of a ternary Muller C element proposed by Sundararajan et al. [16].

4.1 Resistive load ternary inverter

Figure 2 shows the circuit diagram of a resistive load ternary inverter. The circuit consists of two N-channel transistors and two resistors. CNTFET with two different diameters are deployed in this circuit. Transistor T_{LN1} is a low threshold CNTFET and has a diameter of $(19, 0)$ and transistor T_{HN2} is high threshold CNTFET and has a diameter of $(10, 0)$. T_{LN1} and T_{HN2} have threshold voltages of $0.29\ V$ and $0.59\ V$ respectively. The resistors used in this circuit are both $100\ K\Omega$. The circuit is operated at voltage of $V_{DD} = 0.9V$. **Table 3** details the chirality vector, diameter and the threshold voltage of all transistors in **Figure 2**. If the input signal a is below $300\ mV$, none of the transistors are on and the value at output $\hat{a}$ is

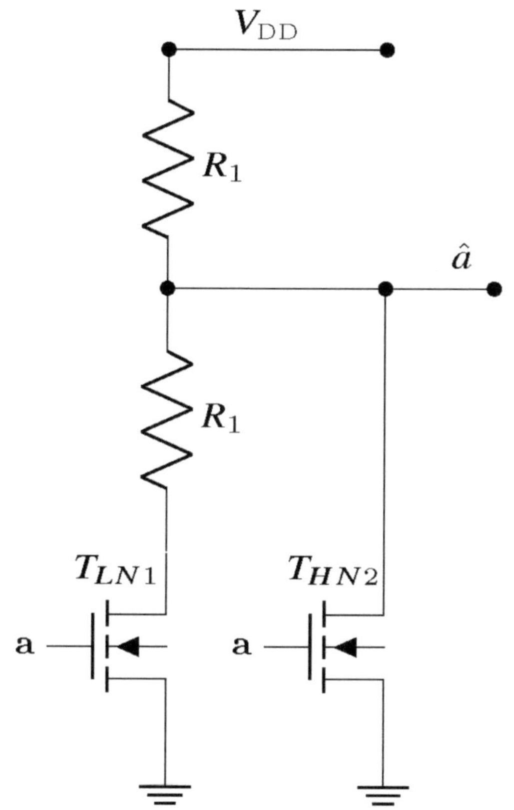

Figure 2.
Resistive CNTFET ternary inverter [6]. Value of R_1 is 100 KΩ.

Transistor	Chirality Vector (n,m)	Diameter(nm)	V_{th}(V)
T_{LN1}	(19,0)	1.487	0.29
T_{HN2}	(10,0)	0.783	0.56

Table 3.
CNTFET chirality, V_{th} and threshold voltages for resistive ternary inverter [7].

0.9 V. If a is operated between 300 mV and 600 mV transistor T_{LN1} turns on, T_{HN2} is off and the circuit operates as a voltage divider and $\hat{a}$ is exactly at half of V_{DD} which is 0.45 V. When a is operated above 600 mV, then the pull down network consisting of both the N channel CNTFETs is on and voltage at $\hat{a}$ is 0 V.

4.2 Static complementary ternary inverter

The problem with the resistive load ternary inverter is that resistors are large, bulky and are prone to noise. Also, static resistors draw leakage currents from the supply and are not suitable for implementation in large scale CNTFET circuits. As an alternative to resistive load ternary inverter, Lin et al. proposed a static complementary version of ternary inverter that employs P-channel and N-channel CNTFETs as shown in **Figure 3**. The resistor in pull up network of the circuit in **Figure 2** is replaced with the two P-Channel CNTFET with varying diameters and the voltage divider resistor is replaced with two diode connected complementary CNTFET. The diode connected transistors have a nominal threshold voltage (V_{th})

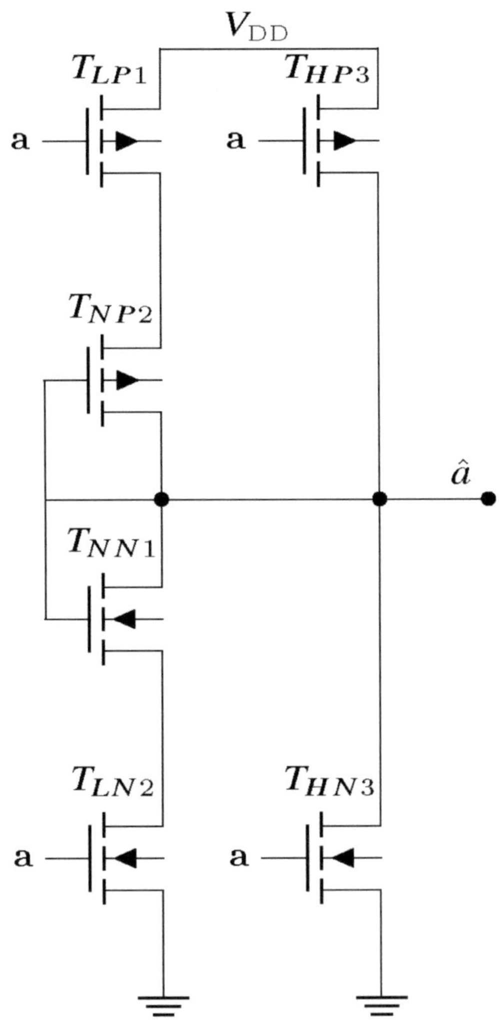

Figure 3.
Static complementary CNTFET ternary inverter [7].

Transistor	Chirality Vector (n,m)	Diameter(nm)	V_{th}(V)
T_{NN1}, T_{NP2}	(13,0)	1.018	0.43
T_{LN2}, T_{LP1}	(19,0)	1.487	0.29
T_{HN3}, T_{HP3}	(10,0)	0.783	0.56

Table 4.
CNTFET chirality, V_{th} and threshold voltages for static ternary inverter [7].

of 0.43 V. **Table 4** describes the chirality values, diameter and threshold voltages of all the transistors in **Figure 3**. If the input signal a is operated below 300 mV, T_{LP1} and T_{HP3} are on and transistors T_{LN2} and T_{HN3} are off. The voltage at $\hat{a}$ is 0.9 V. When input signal a is operated between 300 mV and 600 mV, T_{HP3} is off and diode connected transistors (T_{NP2} and T_{NN1}) are turned on along with T_{LP1} and T_{LN2}. This circuit configuration operates as a voltage divider and produces a voltage drop of 0.45 V at $\hat{a}$. When input signal a is operated above 600 mV, pullup network consisting of all the P-channel transistors is off and all the N-channel transistors are on pulling down voltage at $\hat{a}$ to 0 V.

4.3 Ternary Muller C element

In this section, we will review the circuit implementation of ternary Muller C element described by Sundararajan et al. [16]. Muller C element is a common logic gate that is deployed in asynchronous logic [17, 18]. **Figure 4** shows a circuit schematic and a logic representation of the Muller C element [19]. The basic operation of the C element can be described as follows: When the logic values of inputs a and b are the same, the output c is transparent and input value is latched. Otherwise, output c will retain its previous value. The logic equation describing the behavior of the Muller C element is described as follows [20].

$$c = \hat{c} \cdot (a + b) + a \cdot b \tag{10}$$

where a, b are the two inputs and $\hat{c}$ denotes the previous state of the output c. C element consists of two parts: C-not and S-gates [21]. The C-not part consists of two pmos and two nmos transistors connected in series. The S-gates consists of two cross coupled inverters employing weak feedback and is a widely used implementation proposed by Martin et al. [18]. The ternary Muller C element is similar to its binary counterpart except that the inputs and the output could take three logic values. **Table 5** shows the truth table of ternary Muller C element. **Figure 5** shows the circuit schematic of ternary Muller C element. The C-not portion of the Muller C consists of five P-CNTFETs and five N-CNTFETs. In this design, three kinds of CNTFET transistors having three different chirality vectors are used to realize three different threshold voltages. The chirality vectors used, their corresponding diameters and the resulting threshold voltages are the same as static complementary ternary inverter. The values are listed in **Table 4**. "S" gates consist of the ternary inverters connected in a feedback. To enable correct operation, feedback inverter is a weak inverter with less number of tubes. To realize weak feedback, the number of tubes in all CNTFET transistors in the C-not part and in the strong inverter $I1$ is 12. The number of tubes in weak inverter $I2$ is 3. The inverters are based on a static complementary style discussed in Section 4.2. The operation of the circuit is as follows: When the inputs a and b are below 300 mV P-CNTFET transistors T_{LP1}, T_{LP2}, T_{HP4} and T_{HP5} are on as shown in **Figure 5**. The node C_{out} is pulled to logic 2

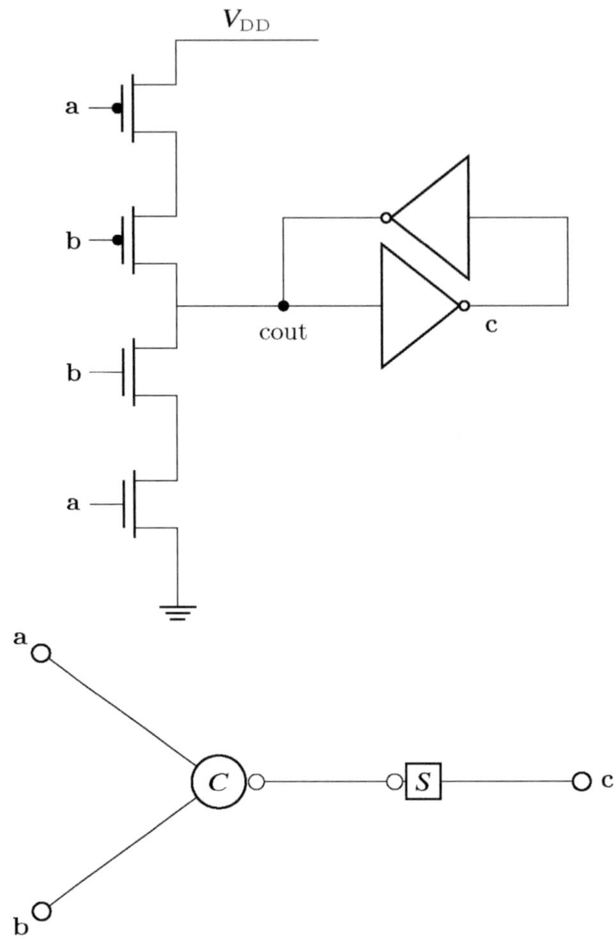

Figure 4.
Binary CMOS Muller C element: Circuit schematic and its logic representation. It has two inputs a, b and an output c. the C element consists of "C-not" and "S" gates [16].

Input a	Input b	Output c
0	0	0
0	1	*S*
0	2	*S*
1	0	*S*
1	1	1
1	2	*S*
2	0	*S*
2	1	*S*
2	2	2

Table 5.
Ternary Muller C-element truth table. S indicates previous state.

and node c is pulled to a low value (logic 0). When the input voltages are raised above 300 mV, transistors T_{HP4}, T_{HP5}, T_{HN4} and T_{HN5} are off as shown in **Figure 6**. Combination of transistors T_{LP1}, T_{LP2}, T_{LN2}, T_{LN3} along with the diode connected

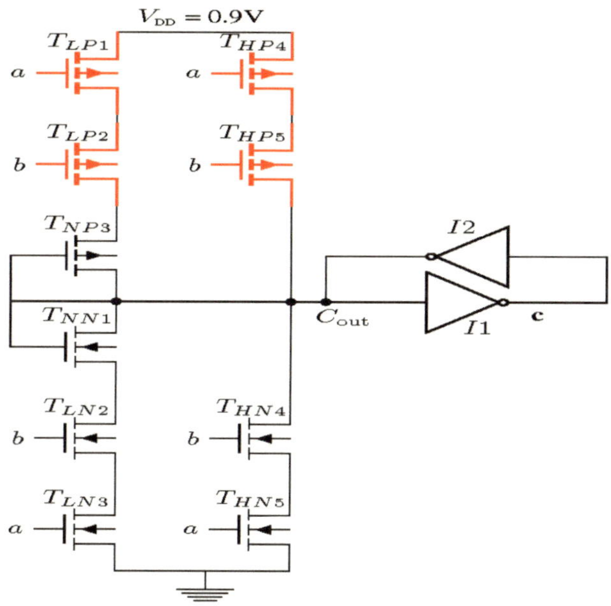

Figure 5.
Schematic for all inputs at logic 0.

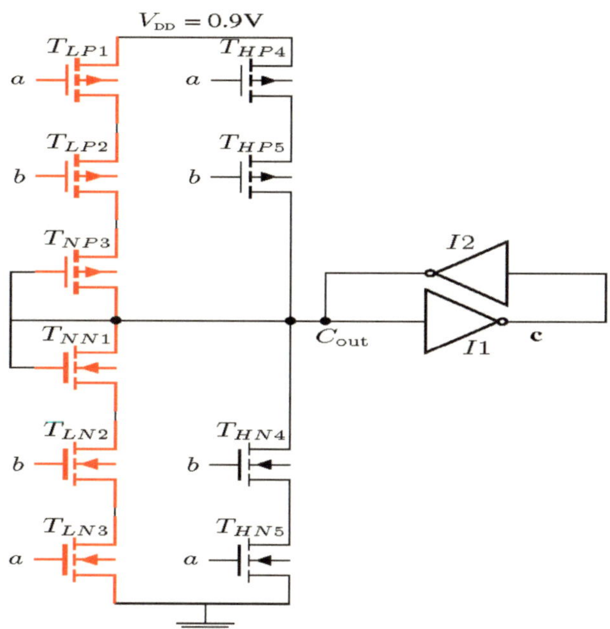

Figure 6.
Schematic for all inputs at logic 1.

transistors T_{NP3} and T_{NN1}, produces a voltage drop of 0.45 V (logic 1) at node C_{out}. This intermediate logic value is fed to the inverter latch, producing a voltage drop of 0.45 V or logic 1 at the output c. As the input voltages are raised above 600 mV, transistors T_{HP5} and T_{HP4} are off and transistors T_{HN4} and T_{HN5} are on as shown in **Figure 7** which pull the node C_{out} to a logic low value and drive node c to a high value (logic 2). The ternary C element was built and evaluated using Synopsys HSPICE circuit simulator. The design was simulated using CNTFET parameter

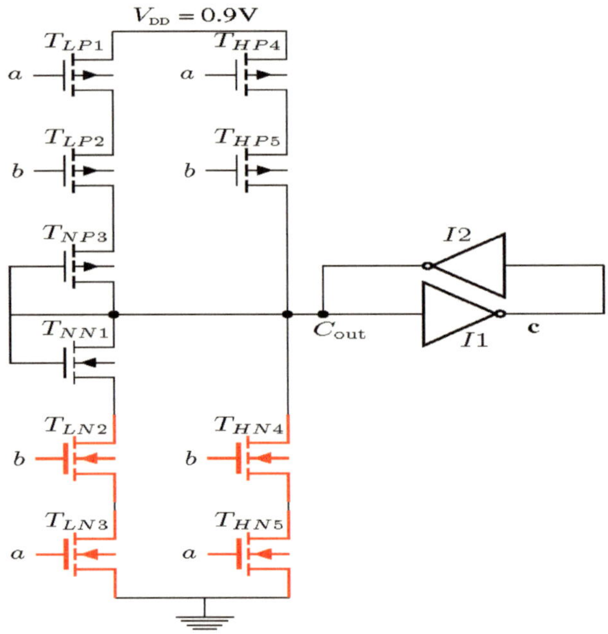

Figure 7.
Schematic for all inputs at logic 2.

models provided by Stanford Nano-electronics Group in the 32 *nm* technology node [22]. The operating supply voltage was 0.9 *V* and the temperature was 27°C. **Figure 8** shows the transient simulation result [16]. In this simulation, input *a* is held at a steady state and input *b* is swept from logic low to a high value. The output changes state when the two inputs are at the same level of logic. When inputs *a* and *b* are at a different logic level, the output holds the previously latched logic value. The inputs are swept from logic low to logic high and then changed to logic low in constant steps and output is observed for a full low to high and a high to low transition. The propagation delay is the average of four terms: the delay from logic 0

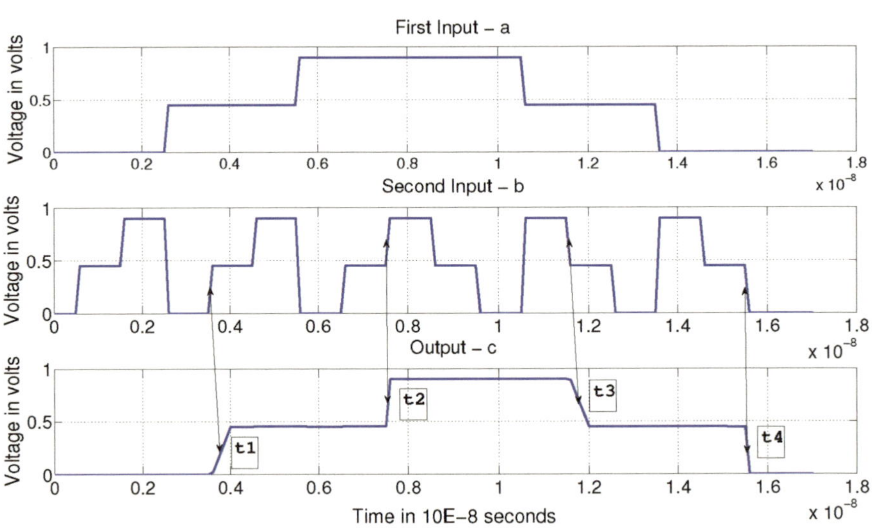

Figure 8.
Transient results of static complementary Muller C [16].

to logic 1 ($t1$), the delay from logic 1 to logic 2 ($t2$), the delay from logic 2 to logic 1 ($t3$), and the delay from logic 1 to logic 0 ($t4$).

5. Variations in CNTFET devices

This section describes the process variation in CNTFET technology. Process variation (PV), an artifact of aggressive technology scaling, causes uncertainty in integrated circuit characteristics. Imperfect lithography, doping fluctuations and imperfect planarization are some of the causes of variation that strongly affect the channel length, oxide thickness, width and eventually the threshold voltage of CMOS transistors. Such variation leads to unpredictable circuit behavior. Due to the random nature of manufacturing process, various effects such as ion implantation, diffusion and thermal annealing have induced significant fluctuations of the electrical characteristics in nanoscale CMOS [23]. CNTFETs are also affected by manufacturing variation caused by imperfect fabrication. CNTFETs not only suffer from traditional process variations, which are in common with the CMOS technologies but they also have their unique source of variations. Paul et al. showed that the CNTFET devices are significantly less sensitive to stochastic variations such as process-induced variations due to their inherent device structures and geometric properties [24]. In addition, CNTFETs are subject to CNT specific variation sources including: CNT type variations, CNT density variations, CNT diameter variations, CNT alignment variations and CNT doping variations [25]. **Figure 9** shows the contributions of each of the aforementioned sources of variations to the on-state current I_{on} in 32 nm technology. To provide sufficient drive current comparable to CMOS, multiple CNTs are deployed in the channel of a single CNTFET. Due to statistical averaging, the impact of the CNT diameter, doping and alignment variations on the on current is significantly reduced for CNTFET with multiple CNTs in the channel [26]. Hence, significant on state current variations are caused by metallic CNT induced count variations and CNT density variations. To overcome the effects of count variations due to metallic CNTs a technique called VLSI-compatible Metallic CNT Removal (VMR) has

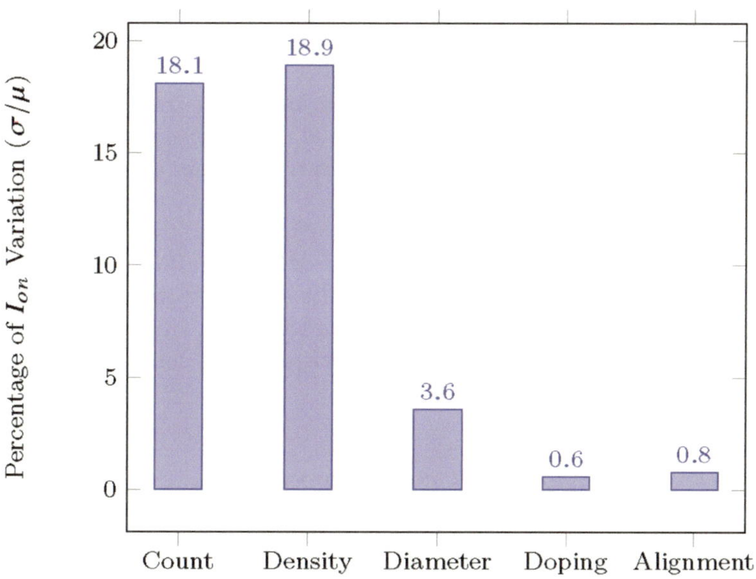

Figure 9.
Contributions of various CNT specific variation sources [25].

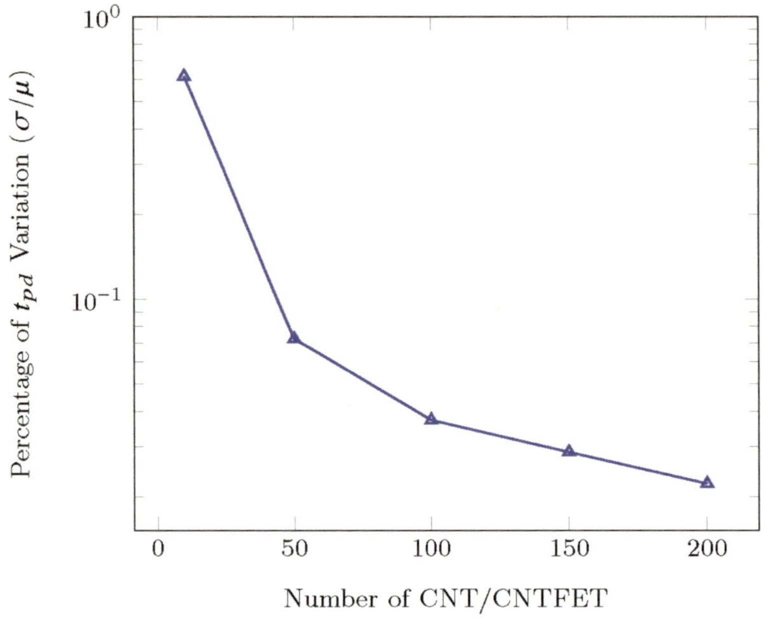

Figure 10.
Percentage of t_{pd} variation for the ternary Muller C element.

been proposed that combines new CNTFET circuit design techniques with the processing [27]. Asymmetrically-Correlated CNTs (ACCNT) is another imperfection immune design technique to tolerate the metallic CNTs [28]. Therefore, CNT density variation is the dominant source of variation in CNTFETs.

5.1 Analysis of process variations

As discussed in Section 5, CNT density is the critical parameter that needs to be accounted for variation aware design. CNT density variation manifests itself as varying amounts of CNT under the gate of every CNTFET transistor. CNT density variation follows a normal distribution with mean $\mu = N$ and the variance $\sigma = \sqrt{(N/2)}$ where N is the expected number of CNTs under the gate of each CNTFET [29]. To assess the impact of CNT density variation on the transient delay, 1000 sample Monte Carlo simulations were performed on the proposed ternary C element. **Figure 10** shows the percentage variation of transient delay with the variation in the number of CNTs per CNTFET.

The C element was driving a capacitance of $25fF$. From the plot, we see that when number of CNT in CNTFET is lower, the delay variation is higher and as we increase the number of CNT in the CNTFET, the delay variation goes down. The variation percentage is 0.6 when the number of tubes in a single CNTFET is 10 and it drops $30X$ lower to 0.02 when number of tubes in a single CNTFET is increased to 200.

6. Defect and fault tolerance in CNTFET MVL logic

This section will review a method for fault tolerance in CNTFET Multi-valued logic that was proposed by Sundararajan et al. [16]. The method called Restorative FeedBack (RFB), provides fault resilience against Single Event Upset (SEU) [19]. Triple Modular Redundancy (TMR) is one of the mostly commonly used method

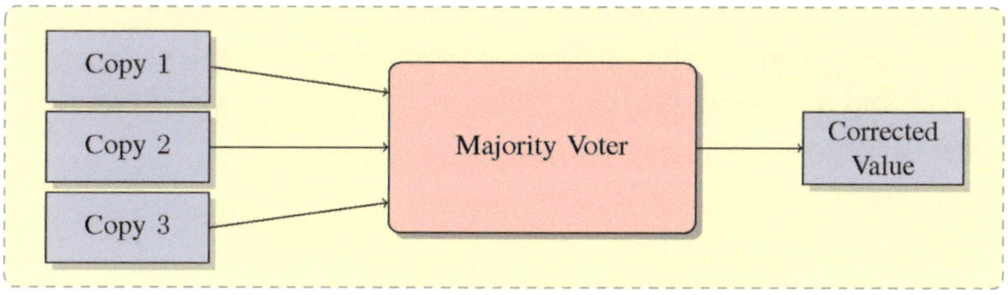

Figure 11.
Triple modular redundancy (TMR).

for fault tolerance in computer systems [30]. In TMR, a logic circuit is replicated three times and a majority vote is taken on the combined outputs. **Figure 11** shows the TMR method, where three copies of a logic circuit are made and a majority vote is taken on those combined copies to obtain a fault free logic value. TMR provides resilience against single error faults. The main drawback of TMR is that logic overhead is high due to replication and and a fault in majority voter can render the technique inefficient. Also, simultaneous error in two copies of the logic value can result in propagation of the error value in the downstream logic. From a MVL perspective, TMR cannot be applied to MVL as it is ambiguous for MVL. As an alternative to TMR, Winstead et al. proposed RFB method, which is an improved version of TMR [19]. RFB method can correct single error faults and CMOS implementations were shown for both binary and ternary logic [19]. RFB method replaces the majority gate in TMR with Muller C elements. The idea of RFB method is based on the inherent fault correcting capabilities of the Muller C element. RFB method is based on two key ideas related to the Muller C element:

- If the C element output is connected back to one of C element's input, then noise in other C element's input will be suppressed every time the C element evaluates.
- Muller C element has inherent fault correcting properties. C element only changes its output state when all inputs have the same logic value. This property can be exploited for fault masking.

Figure 12 shows an example of how a noisy input signal will be rendered less noisy, each time the signal passes through the C element. Suppose ε_a is the error probability associated with input signal a. Then the corresponding Log Likelihood Ratio (LLR), l_a, is defined as [21].

$$l_a = \ln(1 - \varepsilon_a)/\varepsilon_a. \qquad (11)$$

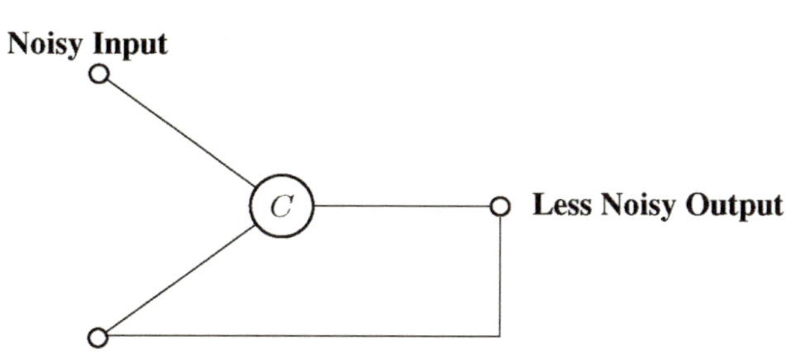

Figure 12.
Muller C element with feedback.

The magnitude of l_a is the confidence that signal a is correct. If, at any given time t, the error events appearing at the C-element's inputs are independent from each other, and also independent from the C-element's internal state, then the output LLR is

$$l_c = l_a + l_b \tag{12}$$

From Eq. (12), we notice that as $l_b \rightarrow l_c, l_a \rightarrow 0$. If the C element's output is connected back to the C-element's input then other input that is noisy is rendered less noisy every time the C-element evaluates the output [19]. **Figure 13** shows the schematic implementation of the RFB method where the C -element's output is connected in a feedback fashion to the adjacent C element's input. The other input to the C element is the logic that is replicated thrice (x_1, x_2 and x_3). To fully realize the fault correcting capabilities of the C element, the RFB circuit is operated in two phases: setup phase and restoration phase. There are three inputs signals x_1, x_2 and x_3. Lets asume that the signal x_3 is corrupted and has an incorrect logic value of 1 while x_1 and x_2 are at logic 2. The RFB operation involves two phases and phase switching is accomplished by using a 2-input multiplexer. The select signal ϕ is at logic 0 during the setup phase. The operation in the setup phase is shown in **Figure 14**. During this phase, inputs are barrel shifted to the adjacent C element's output. The error value in x_3 is shifted to y_2. **Figure 15** shows the operation in restoration phase. During this phase, ϕ is at logic 1. The feedback is deactivated and C-not gate's output is transferred to the actual output. The first C element has its inputs held at logic 2 and that causes y_2 to change its state from logic 1 to logic 2. The transient simulation results are shown in **Figure 16** for all possible combinations of error values [16]. The signal of interest is y_2. The setup and the restoration phases corresponding to y_2 are clearly marked and annotated in **Figure 16**. From the plot, we observe that in the setup phase, the error value in x_3 is transferred to signal y_2 and is then corrected in restoration phase, thereby showing the effectiveness of the RFB technique.

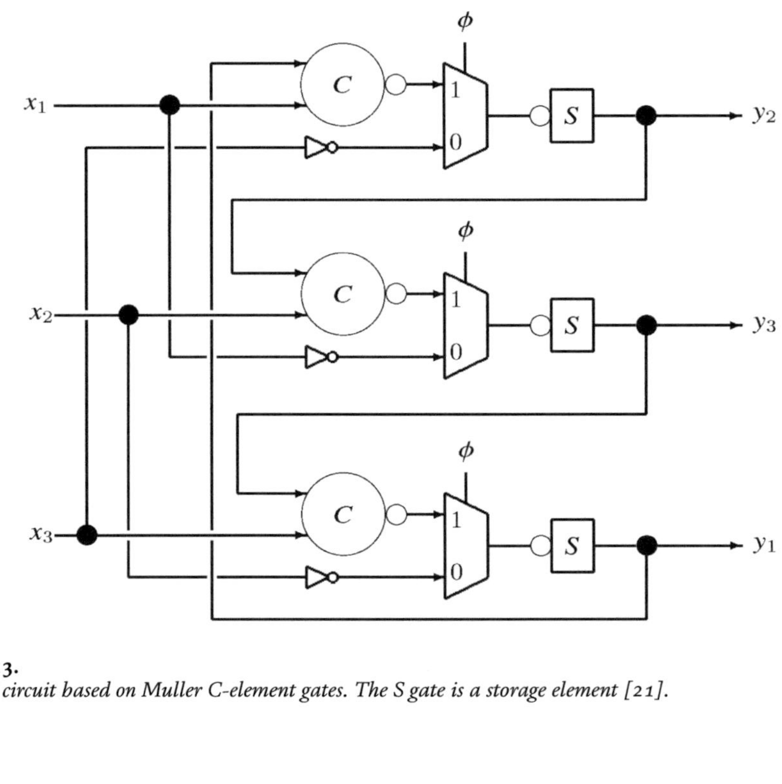

Figure 13.
The RFB circuit based on Muller C-element gates. The S gate is a storage element [21].

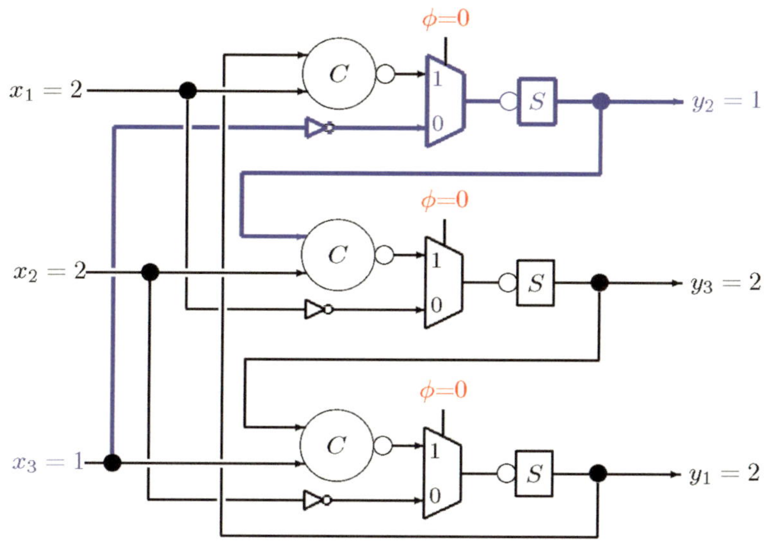

Figure 14.
RFB setup phase.

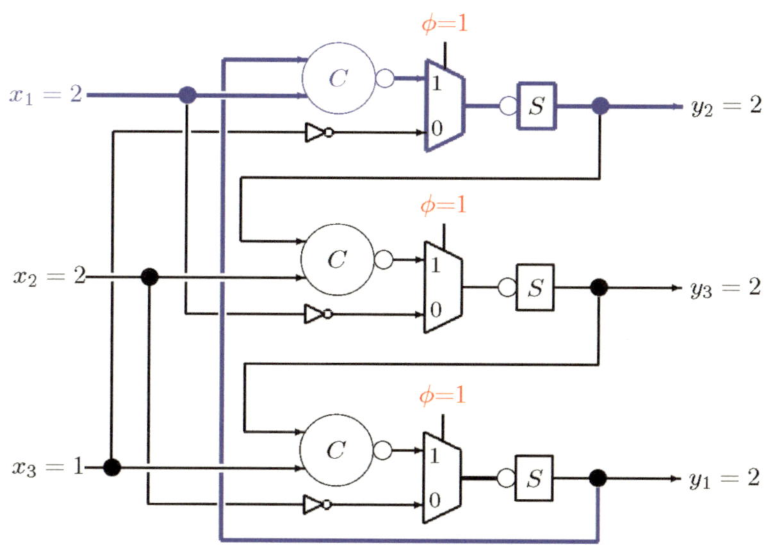

Figure 15.
RFB restoration phase.

7. Conclusion

This chapter discussed a technique for fault tolerance in MVL CNTFET logic. The described technique leverages the error suppression capability of Muller C element to correct single bit error in CNFTET MVL. To realize the technique, a ternary Muller C element is needed. This chapter also discussed the basics of MVL, provided an overview of CNTFET and also discussed the process variation in CNTFET.

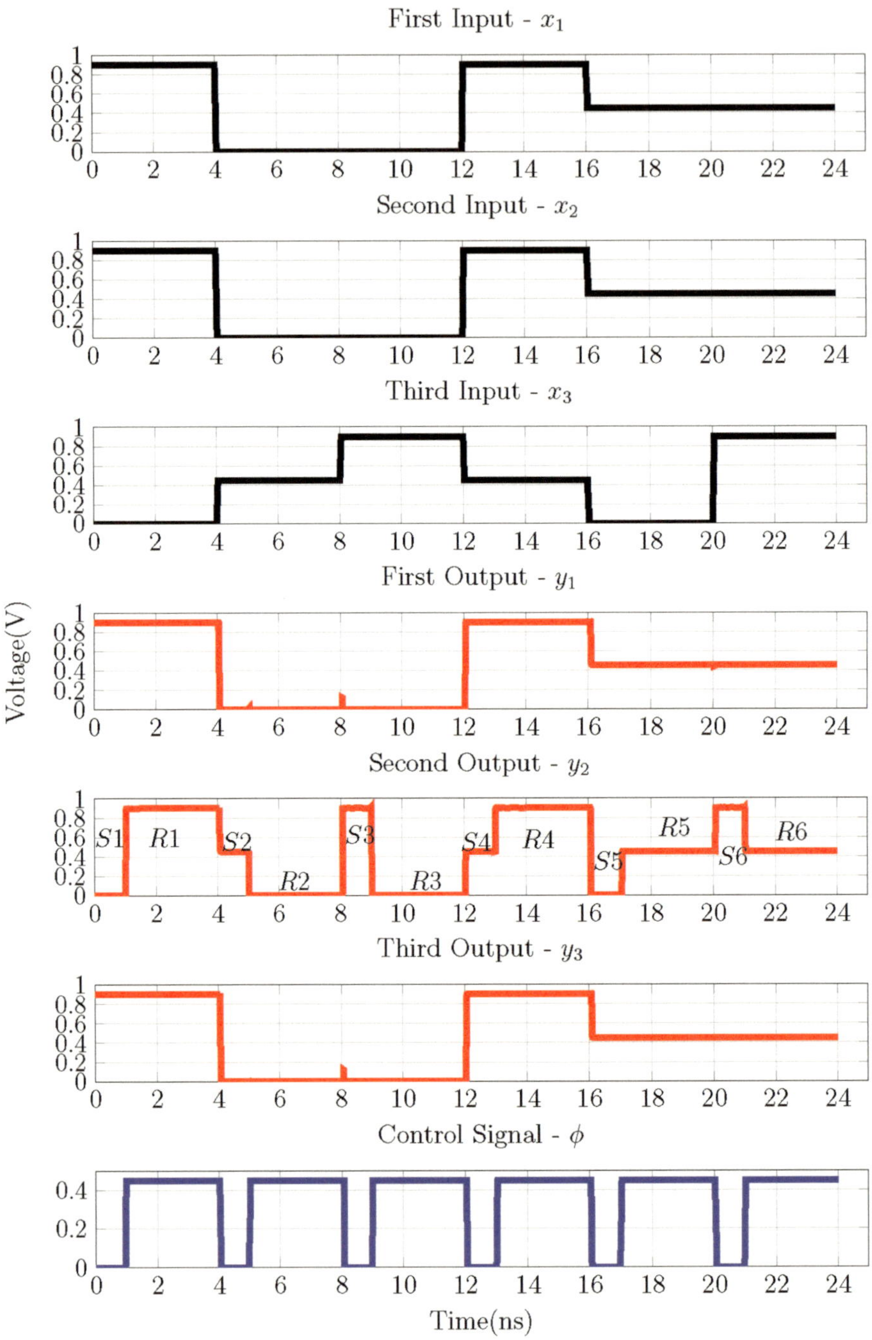

Figure 16.
Transient results of ternary CNTFET RFB method. The input sequence includes the six possible single error patterns. Signals x_1 and x_2 are correct, while x_3 is an error. The error is transferred to y_2 in the set-up phase and is corrected in the restoration phase [16].

Author details

Gopalakrishnan Sundararajan
Intel Corporation, Austin, Texas, USA

*Address all correspondence to: gopal.sundar@aggiemail.usu.edu

References

[1] J. Appenzeller, J. Knoch, R. Martel, V. Derycke, S.J. Wind, and P. Avouris, "Carbon nanotube electronics," *Nanotechnology, IEEE Transactions on*, vol. 1, no. 4, pp. 184–189, dec 2002.

[2] K.K. Likharev, "Single-electron devices and their applications," *Proceedings of the IEEE*, vol. 87, no. 4, pp. 606–632, apr 1999.

[3] S. Onneweer, H. Kerkhoff, and J. Butler, "Structural computer-aided design of current-mode CMOS logic circuits," in *Multiple-Valued Logic, 1988., Proceedings of the Eighteenth International Symposium on*, 0–0 1988, pp. 21–30.

[4] Y. Chang and J.T. Butler, "The design of current mode CMOS multiple-valued circuits," in *Multiple-Valued Logic, 1991., Proceedings of the Twenty-First International Symposium on*, may 1991, pp. 130–138.

[5] Weste Neil H. E. and David Harris, *CMOS VLSI Design: A circuit and systems perspective*, Pearson, 3/e edition, 2006.

[6] A. Raychowdhury and K. Roy, "Carbon-nanotube-based voltage-mode multiple-valued logic design," *Nanotechnology, IEEE Transactions on*, vol. 4, no. 2, pp. 168–179, march 2005.

[7] Sheng Lin, Yong-Bin Kim, and Fabrizio Lombardi, "Design of a ternary memory cell using cntfets," *Nanotechnology, IEEE Transactions on*, vol. 11, no. 5, pp. 1019–1025, Sept 2012.

[8] J. Zhang, A. Lin, N. Patil, H. Wei, L. Wei, H. . P. Wong, and *S. mitra*, "Carbon nanotube robust digital vlsi," *IEEE Transactions on Computer-Aided Design of Integrated Circuits and Systems*, vol. 31, no. 4, pp. 453–471, 2012.

[9] A. Raychowdhury, A. Keshavarzi, J. Kurtin, V. De, and K. Roy, "Carbon nanotube field-effect transistors for high-performance digital circuits–dc analysis and modeling toward optimum transistor structure," *IEEE Transactions on Electron Devices*, vol. 53, no. 11, pp. 2711–2717, 2006.

[10] J. Deng and H. . P. Wong, "A compact spice model for carbon-nanotube field-effect transistors including nonidealities and its application–part i: Model of the intrinsic channel region," *IEEE Transactions on Electron Devices*, vol. 54, no. 12, pp. 3186–3194, 2007.

[11] J. Appenzeller, "Carbon nanotubes for high-performance electronics: Progress and prospect," *Proceedings of the IEEE*, vol. 96, no. 2, pp. 201–211, feb. 2008.

[12] Jie Deng and H-S.P. Wong, "Modeling and analysis of planar-gate electrostatic capacitance of 1-D FET with multiple cylindrical conducting channels," *Electron Devices, IEEE Transactions on*, vol. 54, no. 9, pp. 2377–2385, sept. 2007.

[13] Sheng Lin, Yong-Bin Kim, and F. Lombardi, "CNTFET-based design of ternary logic gates and arithmetic circuits," *Nanotechnology, IEEE Transactions on*, vol. 10, no. 2, pp. 217–225, march 2011.

[14] Bo Wang, C. H. Patrick Poa, Li Wei, Lain-Jong Li, Yanhui Yang, and Yuan Chen, "(n,m) selectivity of single-walled carbon nanotubes by different carbon precursors on coâˆ'mo catalysts," *Journal of the American Chemical Society*, vol. 129, no. 29, pp. 9014–9019, 2007, PMID: 17602623.

[15] A. Lin, N. Patil, K. Ryu, A. Badmaev, L. Gomez De Arco, C. Zhou, *S. mitra*, and H. . Philip Wong, "Threshold voltage and on–off ratio tuning for multiple-tube carbon nanotube fets," *IEEE Transactions on Nanotechnology*, vol. 8, no. 1, pp. 4–9, 2009.

[16] G. Sundararajan and C. Winstead, "Cntfet-rfb: An error correction implementation for multi-valued cntfet logic," in *2016 IEEE 46th International Symposium on Multiple-Valued Logic (ISMVL)*, 2016, pp. 11–16.

[17] M. Shams, J.C. Ebergen, and M.I. Elmasry, "Modeling and comparing CMOS implementations of the C-element," *Very Large Scale Integration (VLSI) Systems, IEEE Transactions on*, vol. 6, no. 4, pp. 563–567, dec. 1998.

[18] Alain Martin, Andrew Lines, Rajit Manohar, Mika Nystroem, Paul Penzes, Robert Southworth, Uri Cummings, and Tak Lee, "The design of an asynchronous mips r3000 processor," 02 1997.

[19] Chris Winstead, Yi Luo, Eduardo Monzon, and Abiezer Tejeda, "Error correction via restorative feedback in *M* -ary logic circuits," *Journal of Multiple Valued Logic and Soft Computing*, vol. 23, no. 3–4, pp. 337–363, 2014.

[20] M. Shams, J.C. Ebergen, and M.I. Elmasry, "A comparison of CMOS implementations of an asynchronous circuits primitive: the C-element," in *Low Power Electronics and Design, 1996., International Symposium on*, aug 1996, pp. 93–96.

[21] Chris Winstead, Yi Luo, Eduardo Monzon, and Abiezer Tejeda, "An error correction method for binary and multiple-valued logic," *Multiple-Valued Logic, IEEE International Symposium on*, vol. 0, pp. 105–110, 2011.

[22] Stanford, "Stanford university CNFET model website [online, accessed 2013]. available: http://nano.stanford.edu/model.php?id=23,"

[23] K. Agarwal and S. Nassif, "Characterizing process variation in nanometer CMOS," in *Design Automation Conference, 2007. DAC '07. 44th ACM/IEEE*, 2007, pp. 396–399.

[24] B.C. Paul, S. Fujita, M. Okajima, T. Lee, H.S.P. Wong, and Y. Nishi, "Impact of process variation on nanowire and nanotube device performance," in *Device Research Conference, 2007 65th Annual*, june 2007, pp. 269–270.

[25] Jie Zhang, N. Patil, H.-S.P. Wong, and *S. mitra*, "Overcoming carbon nanotube variations through co-optimized technology and circuit design," in *Electron Devices Meeting (IEDM), 2011 IEEE International*, dec. 2011, pp. 4.6.1–4.6.4.

[26] Jie Zhang, N.P. Patil, A. Hazeghi, H.-S.P. Wong, and *S. mitra*, "Characterization and design of logic circuits in the presence of carbon nanotube density variations," *Computer-Aided Design of Integrated Circuits and Systems, IEEE Transactions on*, vol. 30, no. 8, pp. 1103–1113, aug. 2011.

[27] N. Patil, A. Lin, Jie Zhang, Hai Wei, K. Anderson, H.-S.P. Wong, and S Mitra, "VMR: Vlsi-compatible metallic carbon nanotube removal for imperfection-immune cascaded multi-stage digital logic circuits using carbon nanotube fets," in *Electron Devices Meeting (IEDM), 2009 IEEE International*, 2009, pp. 1–4.

[28] A. Lin, N. Patil, Hai Wei, *S. mitra*, and H.-S.P. Wong, "A metallic-cnt-tolerant carbon nanotube technology using asymmetrically-correlated cnts (ACCNT)," in *VLSI Technology, 2009 Symposium on*, june 2009, pp. 182–183.

[29] A.A.M. Shahi, P. Zarkesh-Ha, and M. Elahi, "Comparison of variations in MOSFET versus CNFET in Gigascale integrated systems," in *Quality Electronic Design (ISQED), 2012 13th International Symposium on*, 2012, pp. 378–383.

[30] R. E. Lyons and W. Vanderkulk, "The use of triple-modular redundancy to improve computer reliability," *IBM J. Res. Dev.*, vol. 6, pp. 200–209, 1962.